D1570051

*Methods of
Experimental Physics*

VOLUME 25

GEOMETRICAL AND INSTRUMENTAL OPTICS

METHODS OF EXPERIMENTAL PHYSICS

Robert Celotta and Judah Levine, *Editors-in-Chief*

Founding Editors

L. MARTON
C. MARTON

Volume 25

Geometrical and Instrumental Optics

Edited by

Daniel Malacara

Centro de Investigaciones en Optica
Leon, Gto
Mexico

ACADEMIC PRESS, INC.
Harcourt Brace Jovanovich, Publishers

Boston San Diego New York
Berkeley London Sydney
Tokyo Toronto

Copyright © 1988 by Academic Press, Inc.
All rights reserved.
No part of this publication may be reproduced or
transmitted in any form or by any means, electronic
or mechanical, including photocopy, recording, or
any information storage and retrieval system, without
permission in writing from the publisher.

ACADEMIC PRESS, INC.
1250 Sixth Avenue, San Diego, CA 92101

United Kingdom Edition published by
ACADEMIC PRESS INC. (LONDON) LTD.
24-28 Oval Road, London NW1 7DX

Library of Congress Cataloging-in-Publication Data
Geometrical and instrumental optics.
  (Methods of experimental physics; v. 25)
  Includes bibliographies.
  Includes index.
  1. Optics, Geometrical. 2. Optical instruments.
I. Malacara, Daniel, Date- . II. Series.
QC381.G46   1988   535   87-27046
ISBN 0-12-475970-X

Printed in the United States of America
88 89 90 91   9 8 7 6 5 4 3 2 1

CONTENTS

List of Contributors ix
Preface xi
List of Volumes in Treatise xiii

1. Optics and Optical Methods
by R. E. Hopkins and D. Malacara

1.1. Introduction 1
1.2. Optics: A Continuing Contribution to Science and Engineering 2
1.3. The Current State of the Optical Industry in the United States 2
1.4. Procuring Ready-Made Optics 3
1.5. How to Design and Procure New Optical Components . . 4
1.6. The Future of Optics 5
References 5

2. Geometrical Optics
by R. E. Hopkins

2.1. Types of Optical Systems 7
2.2. Basic Laws 10
2.3. Gaussian Optics 22
2.4. Stops, Pupils, Vignetting, and Telecentricity 37
2.5. Paraxial Ray Summary 39
2.6. Third-Order Aberrations: The Aberration Polynomial . . 40

2.7. Optical Design by Computer 51
Appendix 53
References 57

3. The Components in the Basic Optical Systems
by R. E. HOPKINS

3.1. The Components in the State-of-the-Art Optical Systems . 59
3.2. The Terrestrial Telescope 59
3.3. Microscope Optics 67
3.4. Camera Lenses 74
3.5. Projection Lenses and Condensers 99
3.6. Lens Systems for Laser Systems 106
References 118

4. Basic Optical Instruments
by DOUGLAS S. GOODMAN

4.1. Introduction 119
4.2. The Visual System 120
4.3. Magnifiers and Eyepieces 127
4.4. Afocal Systems 132
4.5. The Autocollimator 142
4.6. Projectors 143
4.7. The Microscope 158
4.8. Schlieren Instruments 175
4.9. The Ellipsometer 179
4.10. Refractometers 183
4.11. Spectroscopic Instruments 190
References 228

5. Light Sources
by ZACARIAS MALACARA H. AND ARQUIMEDES MORALES R.

5.1. Introduction 239

5.2. Filament Sources 239

5.3. Arc and Discharge Sources 247

5.4. Light Emitting Diodes 254

5.5. Laser Sources 258

References 270

6. Optical Filters
by FRED GOLDSTEIN

6.1. Introduction 273

6.2. Multilayer Thin Films-Matrix Theory 274

6.3. Selecting Filters 280

6.4. Selecting Filters: Secondary Criteria 297

6.5. Measuring Filter Performance 298

6.6. Vacuum Deposition Techniques 299

6.7. Discussion and Conclusion 301

References 301

CONTRIBUTORS

Numbers in parentheses indicate pages on which the authors contributions begin.

FRED GOLDSTEIN (273), *FTG Software, P.O. Box 358, Chatham, New Jersey 07928*

DOUGLAS GOODMAN (119), *IBM T. J. Watson Research Center, Box 218, Yorktown Heights, New York 10598*

ROBERT E. HOPKINS (1, 7, 59), *Optizon Corporation, 300 East River Rd., Rochester, New York 14623*

DANIEL MALACARA (1), *Centro de Investigaciones en Optica, Apartado Postal 948, 37000 Leon, Gto., Mexico*

ZACARIAS MALACARA H. (239), *Centro de Investigaciones en Optica, Apartado Postal 948, 37000 Leon, Gto., Mexico*

ARQUIMEDES MORALES R. (239), *Centro de Investigaciones en Optica, Apartado Postal 948, 37000 Leon, Gto., Mexico*

PREFACE

Two books covering the field of modern optics have been prepared in this series "Methods of Experimental Physics", separating the material into two parts, one with the title "Geometrical and Instrumental Optics", and the other with the title "Physical Optics and Light Measurements".

The purpose of these books is to help the scientist or engineer who is not a specialist in optics to understand the main principles involved in optical instrumentation and experimental optics.

Our main intent is to provide the reader with some of the interdisciplinary understanding that is so essential in modern instrument design, development, and manufacture. Coherent optical processing and holography are also considered, since they play a very important role in contemporary optical instrumentation. Radiometry, detectors, and charge coupled imaging devices are also described in these volumes, because of their great practical importance in modern optics. Basic and theoretical optics, like laser physics, non linear optics and spectroscopy are not described, however, because they are not normally considered relevant to optical instrumentation.

In this volume, "Geometrical and Instrumental Optics", Chapter One provides a general view of the field of practical optics and mentions some of its most common problems and solutions. Chapters Two to Four consider geometrical optics, optical components, lens design, and optical instruments. Light sources, coherent and noncoherent, which play an extremely important role in optical instruments, are described in Chapter Five. Finally, optical filters and some practical computation methods for them are described in Chapter Six.

There might be some overlapping of topics covered in different chapters, but this is desirable, since the points of view of different authors, treating different subjects, may be quite instructive and useful for a better understanding of the material.

This book has been the result of the efforts of many people. Professor H. W. Palmer started this project and spent many fruitful hours on it. Unfortunately, he did not have the time to finish his editorial work due to previous important commitments.

I would like to express my great appreciation of and thanks to Professor Palmer and all of the authors, without whom this book could never have been finished. I also thank Dr. R. E. Hopkins and many friends and colleagues for their help and encouragement. Finally, I appreciate the great understanding of my family, mainly my wife Isabel, for the many hours taken away from them during the preparation of these books.

DANIEL MALACARA
Leon, Gto. Mexico.

METHODS OF EXPERIMENTAL PHYSICS

Editors-in-Chief
Robert Celotta and Judah Levine

Volume 1. Classical Methods
Edited by Immanuel Estermann

Volume 2. Electronic Methods, Second Edition (in two parts)
Edited by E. Bleuler and R. O. Haxby

Volume 3. Molecular Physics, Second Edition (in two parts)
Edited by Dudley Williams

Volume 4. Atomic and Electron Physics—Part A: Atomic Sources and Detectors; Part B: Free Atoms
Edited by Vernon W. Hughes and Howard L. Schultz

Volume 5. Nuclear Physics (in two parts)
Edited by Luke C. L. Yuan and Chien-Shiung Wu

Volume 6. Solid State Physics—Part A: Preparation, Structure, Mechanical and Thermal Properties; Part B: Electrical, Magnetic, and Optical Properties
Edited by K. Lark-Horovitz and Vivian A. Johnson

Volume 7. Atomic and Electron Physics—Atomic Interactions (in two parts)
Edited by Benjamin Bederson and Wade L. Fite

Volume 8. Problems and Solutions for Students
Edited by L. Marton and W. F. Hornyak

Volume 9. Plasma Physics (in two parts)
Edited by Hans R. Griem and Ralph H. Lovberg

Volume 10. Physical Principles of Far-Infrared Radiation
By L. C. Robinson

Volume 11. Solid State Physics
Edited by R. V. Coleman

Volume 12. Astrophysics—Part A: Optical and Infrared Astronomy
Edited by N. Carleton
Part B: Radio Telescopes; Part C: Radio Observations
Edited by M. L. Meeks

Volume 13. Spectroscopy (in two parts)
Edited by Dudley Williams

Volume 14. Vacuum Physics and Technology
Edited by G. L. Weissler and R. W. Carlson

Volume 15. Quantum Electronics (in two parts)
Edited by C. L. Tang

Volume 16. Polymers—Part A: Molecular Structure and Dynamics; Part B: Crystal Structure and Morphology; Part C: Physical Properties
Edited by R. A. Fava

Volume 17. Accelerators in Atomic Physics
Edited by P. Richard

Volume 18. Fluid Dynamics (in two parts)
Edited by R. J. Emrich

Volume 19. Ultrasonics
Edited by Peter D. Edmonds

Volume 20. Biophysics
Edited by Gerald Ehrenstein and Harold Lecar

Volume 21. Solid State: Nuclear Methods
Edited by J. N. Mundy, S. J. Rothman, M. J. Fluss, and L. C. Smedskjaer

Volume 22. Solid State Physics: Surfaces
Edited by Robert L. Park and Max G. Lagally

Volume 23. Neutron Scattering (in three parts)
Edited by K. Sköld and D. L. Price

Volume 24. Geophysics—Part A: Laboratory Measurements; Part B: Field Measurements
Edited by C. G. Sammis and T. L. Henyey

Volume 25. Geometrical and Instrumental Optics
Edited by Daniel Malacara

1. OPTICS AND OPTICAL METHODS

R. E. Hopkins

Optizon Corporation
300 East River Rd
Rochester, New York 14623

D. Malacara

Centro de Investigaciones en Optica, A. C.
Apdo. Postal 948
37000 Leon, Gto. Mexico

1.1. Introduction

Optics is generally divided into two areas: geometrical optics and physical optics. To the extent that it is fair to make a general assignment, geometrical optics has been assigned to optical engineering, while physical optics has remained in the realm of physics. The division is not clear, nor is it a universal constant. The content of the physics area creeps over into the engineering area as soon as research introduces a useful and needed concept; and the rapidity with which new concepts and products are generated is dictated largely by new demands on the engineering area. There is no escape from the fundamental relationship between the two areas: the limitations on geometrical optics, regardless of advances in computer design and application, ingenuity, and hope are, and always will be, the limits imposed by physical optics.

Electronics and optics, interesting fields in themselves, may be regarded primarily as tools for the researcher. The preponderance of knowledge gained in all of science and engineering has been gained through use of the electromagnetic spectrum. In one sense, optics is indeed the "queen of the battle;" but in a more realistic sense, optics must accept the utilitarian role of being a servant to those who fight the battle. It is not a reduction in status; rather, it is an extension occasioned by the fact that optics provides the most valuable of all observational techniques.

Optics has been the basis of experimental observation from the time that Adam looked at Eve. Scientific and technological progress required new and more refined optical instruments; they required new applications of old ideas every bit as much as they required new techniques and devices to solve both new problems and old problems with more accuracy than heretofore possible. More resolution, better accuracy is always the goal, but a goal that falls short of perfection because of the ultimate limitations of physical optics.

1.2. Optics: A Continuing Contribution to Science and Engineering

Since 1960 there have been several new developments that have put physical optics onto center stage in both science and engineering. The ability to generate coherent light sources, the incredible progress in computing equipment, the ability to make new detectors on microchips, and the development of fiber optics have led to outstanding technical advances which are rapidly requiring engineering development. Many of the developments, as always in the past, will depend on the classical optical components such as lenses, mirrors, and prisms.

It must be granted that optical instrumentation has been a cornerstone for research, and a fundamental tool in technological progress. Nearly every new technology requires new optical configurations. Modern communications, microelectronics, laser applications, and business machines are all seeking new optical devices. As new optical materials are developed, they, too, stimulate further applications for optical elements.

The intent in this book is to give an up-to-date presentation of the state of the art in optical science and engineering.[1-10] It will acquaint the reader with some of the tried and true methods of design and show how to effectively make use of ready-made components in new systems.

1.3. The Current State of the Optical Industry in the United States

In spite of the widespread use of optics in science and technology, a strong industry has never developed in America. This is in contrast to the industries developed around mechanical and electrical engineering. There is a "rule of thumb" that says that the engineering in an optical device is 10% optical and 90% electrical and mechanical. The optical percentage, if anything, is diminishing as electronics is taking over even larger roles in instrumentation. This is partly why optical engineering has not had widespread development as a profession. A second reason is that optical devices are usually passive. Passive elements, if properly designed, built, and cared for, seldom, if ever, need replacement.

The United States manufacturers of classical components (lenses, prisms, mirrors, filters) are mostly small companies that have a specialty. A company that manufactures lenses has equipment for making small lenses and usually cannot make large lenses. Equipment for making cylindrical lenses is different from that for spherical optics. Prisms and flats require still another type of equipment.[11,12] There are few companies that can afford all the equipment which is suitable for efficient manufacturing of the whole range of optics required in modern technology. The result is that a prototype will probably have to be made in a small speciality shop. Unfortunately, this is a debatable issue. If the new mechanical, electro-, optical (MEO) device leads to large production, it probably will be sent overseas for production. The overseas company will have the production equipment which can use low-skill workers and can manufacture at much lower labor costs than we can.

1.4. Procuring Ready-Made Optics

The absence of a strong industry devoted to optics has resulted in the absence of a recognized field of optical engineering. Optical systems[13-15] are usually conceived by mechanical or electrical engineers, or by physicists who temporarily find themselves to be "do-it-yourself" optical engineers. They often approach the supposedly ancient and simple subject with a degree of contempt. Unfortunately, this is not conducive to sound systems design, and the results often are serious delays and cost overruns.

The "do-it-yourselfer" should not be too hasty to prejudge this subject as lacking sophistication. In reality, geometrical optics, instrument development and manufacturing[16] are fine-tuned to the economics of the time. Engineers from other disciplines are usually shocked when they visit a typical optical shop and see the ancient procedures being used. These methods actually are superbly matched to the economics of the free market, which usually means small production runs. When one visits a company making millions of elements in a single production run, one will see the most advanced techniques known to the optical and machine tool industries.

The fragmentation of our optical capability means that a designer of new equipment which will require a prototype model should first look for "off-the-shelf" components. There are now several companies (Ealing, Melles Griot, Royln, Oriel, Klinger) that catalogue special types of components.

1.5. How to Design and Procure New Optical Components

When the decision is made to procure off-the-shelf items, one must understand the basics of lens types and have reasonable appreciation of what one can expect to find commercially. One rule to follow is that the lens will not be on a shelf if no one else wants it. The first step is to determine if someone else has need for a similar lens. Is the lens like a photographic lens or is it like an eyepiece? If so, there may be a chance that it is available as an "off-the-shelf" lens. After the lens is found, it must be tested to make sure that it meets the requirements. Herein lies a "booby trap." For some applications the testing may cost as much time and money as to design and build a new lens. Adequately testing[17] a single component is not a trivial task, and unfortunately the houses that supply components do not describe in sufficient detail the performance of the components. The component should be tested under circumstances that are as close to operating conditions as possible. This is seldom possible for the designers developing a new system. When a real bargain component comes into the laboratory and is inserted into the equipment, it forms an image, as most lenses are supposed to do. The usual reaction is excitement with the nice image and the cheap price of the component. Unfortunately, the testing is usually not done as critically as it should have been. As the system becomes more highly developed, the imagery is reviewed with increasing attention, and all too often it is found wanting. Then, this can be a serious problem. Now one knows exactly what the lens does, and a solution may not be possible with any lens, for the problem is boxed in. The rule is: make sure the "off-the-shelf" component performs exactly as required. The time to be critical is in the beginning. Image improvement does not come with aging. If a component is not used in the way it was intended, it will not perform in an optimum way. Tinkering with it by adjusting spaces will not improve it. If it does not meet the requirements, get a different design for the component.

If the decision is to develop a new optical system, a professional designer should be consulted. The day of the powerful computer has spawned many consulting optical designers. The design of a lens system for a prototype should be done with a prototype shop in mind. The design involves knowing the shop's capabilities and reliability, so that the test glass tools can be used and that the tolerancing can be done economically. A formal tolerance analysis may be suitable for a production task but is not appropriate for prototype development. It also does not do any good to specify tight tolerances, unless the buyer is prepared to check to see if the shop can meet the tolerances. A close relationship between the designer and the shop is highly desirable. It becomes costly to procure components without trust in the prototype shop.

Before approaching the designer, it is also important to be sure that the optical problem is thoroughly understood. Experienced designers find that the most common problems they encounter are with customers who believe that they know precisely what they want and present a list of specifications with unrealistic requirements. Often the customer does not appreciate the complexity of the design which is to use an ill-defined light source with an inexpensive condensor, illuminating a precise object imaged onto a special detector surface, then to be sampled, digitized, and image processed. The transfer function from each component to its neighbor is usually not known with the precision comparable to the precision one can calculate by using mathematically convenient assumptions. In a preliminary design phase, the need for first-hand experience is needed. The optical designer must be able to discuss trade-offs with the electrical and mechanical designers, who in turn should also have some understanding of what the optical trade-offs incur.

One of the major requirements of good design is to match the methods of design, fabrication, and testing with the economics of the problem. There never is a design where "money is no constraint." Cost is one of the important considerations for any design, but it is particularly true for optical instrumentation, for one seldom has the opportunity to write off a large development expense against a large production order. In this treatment of optics, the reader will be given some idea of what to expect when searching for an optical solution to a technical problem.

1.6. The Future of Optics

The future of optics is without any doubt very bright. New optical devices and techniques appear every day. They are even replacing and complementing electronic devices with advantage. Lasers and holograms find new applications every day, not only in the laboratory, but also in telecommunications, industry, administration, and even in domestic instruments, like the digital compact disc players. Some day in the near future, the hybrid optical-electronic computer may be a reality.

References

1. R. Kingslake, ed., *Applied Optics and Optical Engineering*, Vol. 1, Academic Press, New York, 1965.
2. R. Kingslake, ed., *Applied Optics and Optical Engineering*, Vol. 2, Academic Press, New York, 1965.
3. R. Kingslake, ed., *Applied Optics and Optical Engineering*, Vol. 3, Academic Press, New York, 1965.

4. R. Kingslake, ed., *Applied Optics and Optical Engineering*, Vol. 4, Academic Press, New York, 1967.
5. R. Kingslake, ed., *Applied Optics and Optical Engineering*, Vol. 5, Academic Press, New York, 1969.
6. R. Kingslake and B. J. Thompson, eds., *Applied Optics and Optical Engineering*, Vol. 6, Academic Press, New York, 1980.
7. R. R. Shannon and J. C. Wyant, eds., *Applied Optics and Optical Engineering*, Vol. 7, Academic Press, New York, 1979.
8. R. R. Shannon and J. C. Wyant, eds., *Applied Optics and Optical Engineering*, Vol. 8, Academic Press, New York, 1980.
9. R. R. Shannon and J. C. Wyant, eds., *Applied Optics and Optical Engineering*, Vol. 9, Academic Press, New York, 1983.
10. A. C. Hardy and F. H. Perrin, *The Principles of Optics*, McGraw-Hill, New York, 1932.
11. F. W. Twyman, *Prism and Lens Making*, 2d ed., Adam Hilger, Bristol, 1952.
12. D. F. Horne, *Optical Production Technology*, Adam Hilger, Bristol, 1972.
13. D. F. Horne, *Optical Instruments and Their Applications*, Adam Hilger, Bristol, 1980.
14. B. K. Johnson, *Optics and Optical Instruments*, Dover Publ., New York, 1960.
15. D. F. Horne, *Lens Mechanism Technology*, Adam Hilger, Bristol, 1975.
16. P. J. Rogers and R. E. Fischer, eds., *Optical System Design, Proc. Soc. Phot. Opt. Instr. Eng.* **339** (1983).
17. D. Malacara, *Optical Shop Testing*, Wiley, New York, 1978.

2. GEOMETRICAL OPTICS

R. E. Hopkins

Optizon Corporation
300 East River Rd
Rochester, New York 14623

2.1. Types of Optical Systems

Optical systems are used to collect light from a variety of sources, transport information, and direct it to a large assortment of detectors. Figure 1 shows a photographic lens which is made up of five elements. All the surfaces are spherical, and the centers of curvature lie on a single axis, i.e., the optical axis. The drawing shows the intersection of the spherical surfaces and the plane of the paper. Any plane containing the optical axis is called the *meridional plane*. The lens shown has rotational axial symmetry about the optical axis.

Figure 2 shows a lens which has a rotational symmetry, but one surface is aspheric. Elements of this type are used in optical systems, but they cost from three to ten times as much as elements with spherical surfaces when they are made of glass. It is difficult to make aspheric surfaces out of glass with the surface smoothness found in spherical lenses. The aspheric surface also has a single axis which, in theory, must pass through the centers of the spherical surfaces in the complete lens. The aspheric element must be carefully mounted. When there is sufficient production, they can be made of plastic at low cost after the molds have been paid for. A word of warning: there are many more optical shops that are willing to make an aspheric surface than can actually do so.

Figure 3 shows a lens system with a decentered element. The axial symmetric lenses shown in Figs. 1 and 2 are theoretical designs. During the manufacturing process, the lens elements are usually decentered to some degree. The effect of these manufacturing errors must be evaluated by designers when the lens is toleranced. There are occasional modern optical systems which are designed to use tilted elements to provide asymmetries in some of the other components. Users beware! If there is no alternative, expect mechanical and testing problems which must be handled with care and at extra expense.

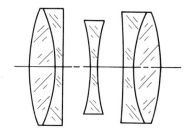

Fig. 1. Photographic lens of five elements.

Optical systems often employ the use of cylindrical and toric surfaces. Spectacle lenses and anamorphic projection lenses are examples.

Figure 4 shows a lens system with a prism which bends the optical axis. In the design phase, the lens is considered as a lens system of four elements on a single axis, with a block of glass between them. In the construction of the system, it is necessary to mount the components on a common axis. This is not a trivial problem for the mirror surface of the prism must be positioned accurately with respect to the optical axis of the two sets of optical components. Folding an optical system like this is hazardous, unless

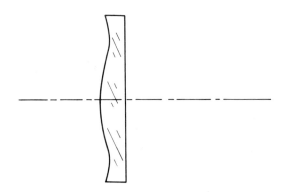

Fig. 2. An aspheric lens element.

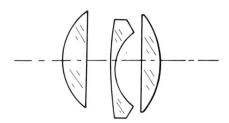

Fig. 3. A decentered lens.

TYPES OF OPTICAL SYSTEMS

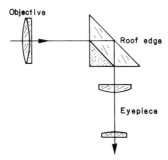

FIG. 4. A system with a prism.

one pays attention to tolerancing, construction, and mounting of the components.

Figure 5 shows a combination of refractive and reflective components. In the design, the elements are considered as rotationally symmetric. The prism is used to provide an erect image and to direct viewing from the object to the observer. Those who have had some optical experience may wish to explain why the prism is a pentaprism. If one has difficulty, one can refer to the literature.[1] The systems shown in Figs. 4 and 5 are simple to design, but they present substantial problems in manufacture. It is necessary to decide whether to provide adjustments to align the components or to use precision during the manufacturing process. Adjustments can move, and fixed machining requires close tolerances on both the optical

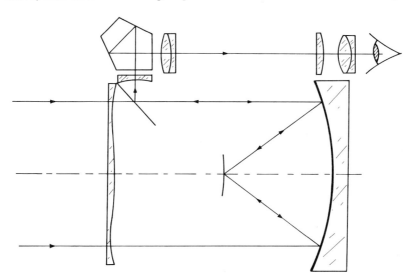

FIG. 5. A metascope. The optics of an infrared viewing telescope.

and mechanical parts. Binoculars, for example, are usually made with some adjustments to keep the prisms aligned; and as a result many binoculars are soon out of alignment. This is particularly true for the less expensive ones. It amounts to false economy, because a pair of misaligned binoculars will not be used for long, and repairing them costs nearly as much as a new pair.

Many systems use gratings or dispersive prisms. Rotating polygons or holograms for laser scanners are becoming common in modern instruments. Quite often the systems are used over a wide spectral range, so mirrors are used rather than refractive elements. Aspheric surfaces are common in mirror systems. Mirror systems are usually something to avoid unless the problem requires large optics or a wide spectral range. Mirror systems have many mounting and obscuration problems. The single surface of a mirror is usually asked to do too much, and therefore it becomes sensitive to mounting.

In spite of the wide variety of lens systems, the vast majority of them turn out to be axially symmetrical systems. This chapter deals with the basic tools needed to evaluate and to design such systems.

2.2. Basic Laws

In this section, the basic laws of geometrical optics and image formation will be described.

2.2.1. The Law of Refraction

The backbone of geometrical optics is ray tracing. The rays of light travel in straight lines as long as the material in which they travel has a homogeneous index of refraction. When the ray encounters an interface between two different materials, refraction of the ray takes place. The equation governing the refraction is written in vector form in Eq. (2.1). The derivation is clear from observing the illustration in Fig. 6.

$$n's' = ns + \Gamma M \qquad (2.1)$$

where

- $\Gamma = n' \cos I' - n \cos I$
- s is a unit vector along the incident ray in the medium to the left of the surface.
- s' is a unit vector along the refracted ray in the medium to the right of the surface.

BASIC LAWS

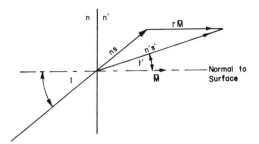

FIG. 6. The vector law of refraction.

- M is a unit vector normal to the surface at the point of incidence; n and n' are the indices of the refraction on the left and right of surface, respectively.

The refraction equation is usually called Snell's law and is written as

$$n \sin I = n' \sin I'. \tag{2.2}$$

It is necessary to add to this equation that the incident, the refracted ray and the normal to the surface all lie in a plane.

In the calculation of optical systems, the object is considered to be made up of a set of point sources of light. The design objective is to shape the lens elements so that each object point is imaged as a perfect point. This means that all the rays passing through the lens from an object point should converge to a single geometrical image point. The rays are normal to the wave fronts of light. The wave front diverging from a source point is spherical, but the rays remain normal to the wave fronts (providing the material is not birefringent). If the designer of the optical system is successful in uniting the rays to a single image point, then the emerging wave front is also spherical and the optical path (OP), from the object to the image point, is equal for every ray. The OP of a ray is the geometrical distance along the ray, multiplied by the index of the material through which the ray is traveling. This does not mean, however, that the optical paths from all object points to their corresponding image points are equal to each other. To repeat: all the ray paths from a single object point to its perfect image are equal.

In most practical systems, the emerging wave front departs from a perfect sphere by an optical path difference (OPD). The OPD represents aberration. When the OPD is less than a wavelength, the distribution of light in the image point is determined by diffraction. OPD values greater than two waves indicate that the light distribution in the image is determined primarily by the geometrical distribution of rays. The OPD is measured with respect to a central chief ray (which is often called the principal ray). When a design calls for a near-diffraction-limited lens, the designer concentrates on reducing the optical-path-length differences to the lowest possible values for several image points over the field of the object. When the images are several waves of OPD, the designer usually concentrates on the geometrical transverse deviations from a perfect point. Most of the modern lens computer programs can calculate both the diffraction integral and what is called the modulation transfer function (MTF). It should be pointed out that most of the design time goes into simply manipulating the design to reduce the geometrical ray deviations or the OPD to small values. The diffraction calculations are only needed near the end of the design to evaluate the precise light distribution in the image. This is why it is stated that ray tracing is the backbone of geometrical optics and optical-instrument design.

2.2.2. The Battle of Notation

A lens system consisting of several components and attributes has many parameters which must be labeled with a systematic convention of signs. The convention used here is widely used, but not universally. The coordinate system is shown in Fig. 7. Figure 8 shows the notation used for the surface parameters: r represents the radius of curvature, t the thickness between surfaces, n the index of refraction, Y the height of a ray on a surface, and U the angle of a ray with respect to the axis. The following rules are used for the notation and the sign of the parameters.

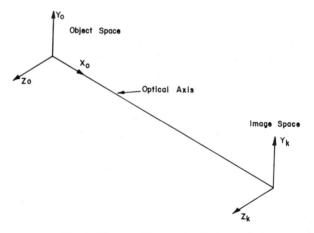

FIG. 7. The coordinate system to be used.

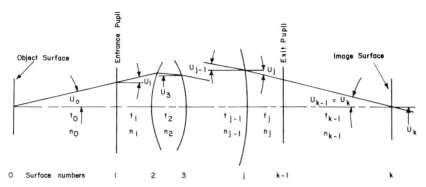

FIG. 8. Notation used to specify surface parameters.

1. The object is assumed to be at one's left, and the light travels from left to right.
2. The object surface (the O surface) is usually considered to be a plane surface. The modern programs, however, will accept a curved object surface.
3. Surface 1 is the entrance pupil plane.
4. Surface j is a general surface in a lens. The subscript j on the j surface is often omitted. For example, r_j is written r while r_{j-1} is written r_{-1}.
5. The radii are positive when the center of curvature of the surface of the lens lies to the right of the pole of the surface, and are negative when the converse is true.
6. The angle U that the ray forms with the axis is positive, in agreement with the convention in analytic geometry. In Fig. 8, U_o is positive, while U_k is negative.
7. The thickness t is positive when the j_{+1} surface lies to the right of the j surface. Thickness t_1 is positive when the front surface of the lens lies to the right of the entrance pupil (as in Fig. 8), and negative in the converse situation. Often t_o is infinite. (With floating-point computers, it is customary to use 10^{10} units to represent infinity.)
8. When the light travels from right to left, the index of refraction is negative, and the signs of the thicknesses are reversed.

Another common notation uses primed quantities to refer to data on the image side of a surface and unprimed quantities for data on the object side of the surface. A comparison between the two notations is given in Fig. 9. When referring to data pertaining entirely to a single surface, it is convenient to use the prime notation for the data on the image side of the surface. This notation will be used here when it is clear that reference is being made to a single surface.

14 GEOMETRICAL OPTICS

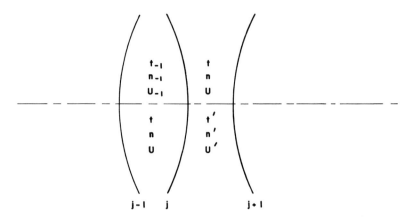

FIG. 9. Comparison between two commonly used conventions for labeling surface data. The lower method will be used in this chapter when it is clearly data pertaining to a single surface.

When deriving formulae for optical calculations, one must develop a systematic way to use the sign convention. An incorrect sign indicates a completely different problem than intended. One must be careful to assign the correct sign for input data, because the computer has no way of knowing what one intends. An incorrect sign can make thousands of calculations and printing worthless.

The following procedures are recommended for deriving optics formulae. (See Longhurst.[2])

1. Draw the diagram accurately.
2. Label all quantities without regard to sign convention. Assume all the data are absolute values.
3. Derive the geometric relations.
4. Reduce the algebra to the equations needed for the calculations, and then insert negative signs in front of any quantity in the diagram which is negative to the agreed-upon convention.

It is not necessary to be concerned about the signs of quantities which do not appear in the final results. Figure 10 shows a doublet with properly assigned input data.

2.2.3. Meridional Ray Tracing

The procedure for deriving equations is illustrated by the derivation of the equations for a meridional ray refracted at a single surface, as in Fig.

Radius of Curvature	Thickness	Index of Refraction
		1.0
5.856	0.9	1.620
-5.670	0.150	1
-5.144	0.600	1.689
-46.729	8.785	1

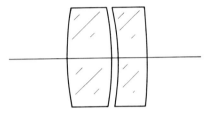

FIG. 10. A doublet objective with properly assigned signs for the curvatures and thicknesses.

11. Notice that all the double arrows on lines indicate lengths and angles. This means that they are labeled as an absolute value. The following geometric relations come directly from the diagram:

$$CA = (L - r) \sin U \tag{2.3}$$

$$\sin I = \frac{CA}{r} \tag{2.4}$$

$$\sin I' = \frac{n}{n'} \sin I, \tag{2.5}$$

$$U' + I' = U + I, \tag{2.6}$$

$$CA' = (L' - r) \sin U'. \tag{2.7}$$

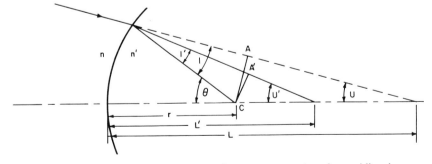

FIG. 11. Construction lines used to derive the ray-trace equations for meridional rays.

In the diagram the U and U' are negative, because the ray is sloping downhill.

Signs for these quantities must then be inserted in the above equations. The correct equations are then:

$$\sin I = \frac{(r-L)}{r} \sin U, \tag{2.8}$$

$$\sin I' = \frac{n}{n'} \sin I, \tag{2.9}$$

$$I' - U' = I - U, \tag{2.10}$$

$$CA' = (r - L') \sin U'. \tag{2.11}$$

The ray is traced to the next surface by use of the transfer equation, as illustrated in Fig. 12. The equation is

$$C_{+1}A_{+1} = CA + (r_{+1} - r + t) \sin U'. \tag{2.12}$$

By successively applying the refraction and transfer equations, a meridional ray may be traced through each surface from the object plane to the image plane. Equations (2.8) to (2.11) were used to show how to keep track of the sign convention. They will be used later to explain the derivation of the paraxial-ray equations. They are simple and easy to use, but they lose accuracy for long radii and small angles of U, because it becomes necessary to subtract two numbers which are nearly equal. This is satisfactory when doing the calculations by hand, and the loss of accuracy can be observed as it happens. When it does happen, a more cumbersome formula can be used. If the program is written for an electronic calculator, a different form of equations should be used. A ray-tracing program for a programmable computer should use a full skew ray trace,[3,4,5] as in the

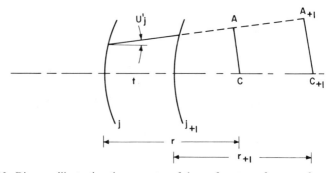

Fig. 12. Diagram illustrating the geometry of the surface-to-surface transfer equation.

Appendix of this chapter. Skew rays are rays that do not remain in the meridional plane as they pass through the system. Further discussion of the equations is not appropriate for this chapter other than to say that extremely sophisticated tracing programs are now deeply imbedded in several commercially available programs. Since this is the innermost loop of an automatic correction program, they are extremely general and fast. On a large-scale computer, thousands of ray surfaces can be traced in a second. It is well to remember, however, that before modern computers, fine lenses were designed by restricting ray tracing to as few as eight meridional rays and close skew rays, called *Coddington's rays*. The meridional rays are the most volatile, so if they and the close skew rays are controlled, the lens is well on the road to good correction.

2.2.4. Paraxial Rays

Paraxial-ray-tracing equations are derived by letting U approach zero as a limit so that $\sin U = U$. With this approximation, the ray-tracing equations reduce to the following single equation, sometimes referred to as the *Gauss equation*:

$$\frac{n}{l} + \frac{n'}{l'} = \frac{n'-n}{r}, \qquad (2.13)$$

where the letter l is used to indicate paraxial data. The equation shows that for an object point there is one image point. The angle U does not enter into the equation. It is useful, however, to think of the paraxial rays as passing through the lens at finite aperture. To do this, one can use the concept of paraxial construction lines (PCL) to represent the paraxial rays. The following substitutions are made in Eq. (2.13):

$$\tan u = \frac{y}{l} \quad \text{and} \quad \tan u' = \frac{-y}{l'}.$$

Equation (2.13) then takes the form

$$n' \tan u' = n \tan u + \frac{(n-n')y}{r}. \qquad (2.14)$$

The geometrical interpretation of this equation is illustrated in Fig. 13. The paraxial rays appear to pass through the lens at finite angles, but they refract at the tangent planes of the optical surfaces.

In a similar way the transfer equation may be written as

$$y_{+1} = y + \frac{t'}{n} n' \tan u'. \qquad (2.15)$$

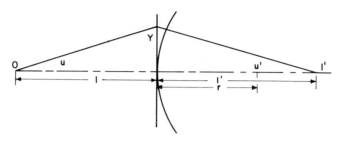

FIG. 13. The representation of paraxial rays which refract on the surface tangent plane.

In order to simplify the paraxial equations, it is customary to replace the tan u with just u. The paraxial equations for refraction and transfer are then

$$n'u' = nu + \frac{y(n-n')}{r}, \qquad (2.16)$$

$$y_{+1} = y + \left(\frac{t'}{n'}\right)n'u'. \qquad (2.17)$$

2.2.5. Graphical Ray Tracing

There is a convenient graphical ray-tracing procedure which is illustrated in Fig. 14. The derivation of this construction is a direct application of Snell's law. The paraxial rays are traced in exactly the same manner as the real meridional rays by using the tangent planes. One can see that the graphical tracing of paraxial rays is an exact method for locating the paraxial images. Figures 15(a) and 15(b) show the graphical tracing of real and paraxial rays through a plano-convex lens. They show that when the lens is turned with the strongly curved surface towards the long conjugate, there

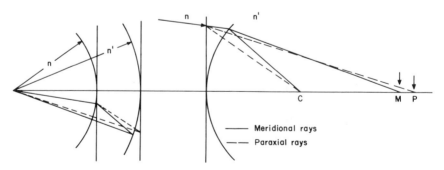

FIG. 14. Graphical construction for tracing paraxial and meridional rays.

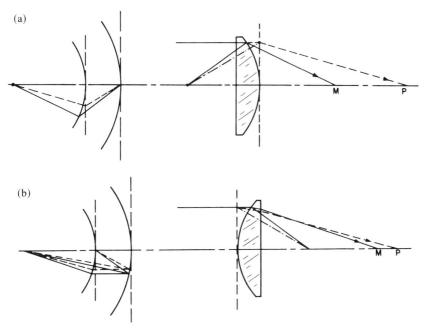

FIG. 15. (a) Construction showing the path of meridional rays in solid lines and paraxial rays in dotted lines. (b) Construction of paraxial and meridional rays through a lens which has been shaped to reduce the spherical aberration.

is reduced spherical aberration. Notice also how closely the paraxial construction lines and the real rays stay together as they pass through the lens. In a well-designed lens, this is nearly true for all the rays. Only when the relative apertures become large do they separate by noticeable amounts. This then is a guide to how the lenses should be shaped and positioned. It also is an explanation of why the paraxial-ray tracing is such a valuable tool for the engineer interested in the mechanical aspects of the problem. The paraxial-ray heights quite accurately indicate the locations of the real rays.

2.2.6. A Paraxial Computation System

Paraxial rays are traced through an optical system using the format shown in Fig. 16. The method is called the *ynu* method. In the example shown, two rays are traced through a two-surface lens in order to locate the focal planes and the cardinal points. This method permits easy calculations for paraxial-ray tracing. The power and versatility of this method becomes

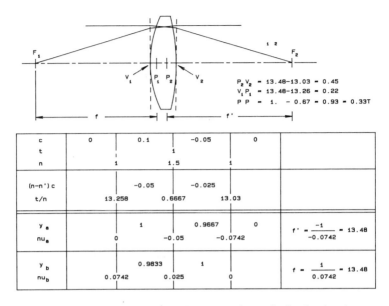

FIG. 16. The computation sheet used to locate the cardinal points in a lens.

evident only after use. Sometimes the rays are traced forwards and sometimes backwards to solve for the curvatures and thickness. It is not necessary to use formulae for the focal lengths of combinations of lenses or equations for the positions of the cardinal points, since all the information lies in the table when the rays have been traced.

The focal length of the lens is

$$f' = \frac{-y_1}{u_k}. \tag{2.18}$$

The back focus is

$$V_2 F_2 = \frac{-y_{k-1}}{u_k}. \tag{2.19}$$

The principal point P_2 is located by means of the equation

$$P_2 V_2 = f' - V_2 F_2. \tag{2.20}$$

The first focal length of the lens is

$$f = \frac{y_{k-1}}{u_1}. \tag{2.21}$$

The front focus distance is

$$F_1 V_1 = \frac{y_1}{u_0}. \tag{2.22}$$

The principal point P_1 is located from the equation

$$V_1 P_1 = f - F_1 V_1. \tag{2.23}$$

The a ray was traced to find the second focal length f', and the b ray was traced backwards to find the first focal length f.

2.2.7. Mirror Systems

Mirror systems with reflecting surfaces are handled by substituting $-n$ for n' on the surfaces where reflection takes place. The system is folded out and ray-traced exactly in the same way as a refracting system. Figure 17 shows the calculation of a double reflection in a single lens. This calculation locates the position of a ghost image. The problem is set up as follows.

1. Draw a picture of the lens and sketch the path of the ray to be traced. One should always do this before tracing a ray. It is worthwhile to develop the ability to accurately sketch the passage of a ray through

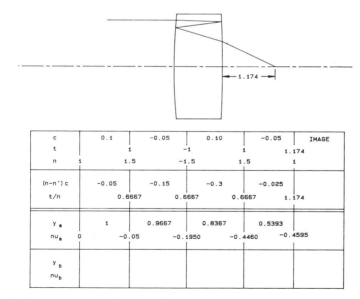

FIG. 17. Tracing rays through a lens with surface reflections in order to locate the position of a reflection ghost.

the lens, for this gives a feeling for how lenses work. It also helps to reveal stupid input mistakes, which computers enjoy grinding on just as much as on the correct problem.
2. Insert the curvature in the order they are encountered by the ray to be traced. Use the sign convention for curvature; do not change the sign on curvatures after a reflection.
3. Insert thicknesses in the order that the ray encounters them. When the ray travels from right to left, insert the thickness as negative. Do the same with the indices as with the thicknesses.

2.3. Gaussian Optics

There is sufficient information from tracing two paraxial rays through an optical system to predict the path of any other ray with a minimum amount of calculation, because there are simple relations between the two rays. The first ray, called the a ray, is labeled using y and u.

The second ray, the b ray, is the oblique ray and referred to by $\bar{y}$ and $\bar{u}$. The following relationship then holds for these two rays on every surface in the system,

$$\bar{y}nu - yn\bar{u} = \bar{y}n'u' - yn'\bar{u}' = H. \tag{2.24}$$

This equation holds for both sides of every surface in the optical system. The H is called the optical invariant. From the data for these two rays, all the Gaussian constants are determined, and the path of any third ray $(\bar{\bar{y}}, \bar{\bar{u}})$ may be computed from a linear combination of the data on the two base rays, using

$$\bar{\bar{y}} = Ay + B\bar{y}, \tag{2.25}$$

$$\bar{\bar{u}} = Au + B\bar{u}. \tag{2.26}$$

The a ray is traced from the base of the object. The b ray is traced from the edge of the object and is directed through the center of the entrance pupil. (See Section 2.4.) These two rays are illustrated in Fig. 18. By using these two rays, the equation for the optical invariant on the object and image surfaces ($y = 0$) reduce to

$$H = \bar{y}_0(nu)_0 = \bar{y}_k(nu)_k. \tag{2.27}$$

This optical invariant is frequently called the *Lagrange equation* and is an indicator of the flux collected by the lens from a uniform Lambertian source. A Lambertian source emits uniform flux into a given solid angle, independent of the angle between the normal to the surface and the direction of viewing. This invariant is also a measure of the information capacity of the

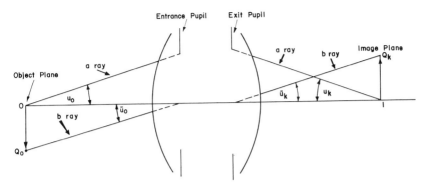

FIG. 18. A diagram illustrating the *a* and *b* rays which are traced through a lens system in order to compute the third-order aberrations.

system. The diameter of the diffraction image of a point source is inversely proportional to $(nu)_k$. The number N of independent diffraction points that the lens can relay from the object to the image is dependent on this value of H, specifically

$$N = \frac{4H^2}{\lambda^2}, \qquad (2.28)$$

where $\lambda = 5 \times 10^{-4}$ mm, $N = 16 \times 10^6 H^2$.

A value of H near unity (mm) means a large number of image points. Table I lists the number of image points that some common types of lenses can image. This large number of image points indicates the remarkable ability of optical systems and shows why they are so useful as information-gathering tools.

TABLE I. The Approximate Number of Diffraction-Limited Spots Imaged by a Selection of Commercial Lenses

	Number of Spots
Microscope Objectives	200,000
Telescope Objectives	170,000
35-mm Camera Lens $F/8$	12,000,000
Aerial Camera Lens 24 inch $F/3.5$	313,000,000
Step-and-Repeat Camera Lens	60,000,000*
Mask Printing Lens	600,000,000
Laser-Scanning Lens	900,000,000*

* These lenses have 50% MTF for each of the image spots.

When paraxial rays are traced, they should be traced through the lens at the height of the full aperture and at the full field coverage. The value of H is then a direct measure of the magnitude of the design problem. It may be simple enough to trace paraxial rays through a lens system with over 10 million image points, but if the lens is to be built, it must have precision surfaces and centering of the elements. One should not be too greedy for a large H without being prepared to face the manufacturing problems.

Many optical systems have H greater than unity, but the image quality degrades near the edges of the field, and there are not as many image points as given by Eq. (2.28). For example, H for a pair of binoculars is around 1.4, but the system is diffraction-limited for no more than $H = 0.2$. The diffraction-limited performance is needed over only a small central region. The remainder of the image field is used for search, detection of motion, and the reduction of a tunnel-vision appearance.

2.3.1. System Magnification and the Abbe Sine Condition

The following expression, obtained from Eq. (2.27), shows that the system magnification m is given by

$$m = \frac{\bar{y}_k}{\bar{y}_0} = \frac{(nu)_0}{(nu)_k}. \tag{2.29}$$

There is also an important generalization for marginal rays, of the Lagrange equation, called the *Abbe sine law*, which states that

$$M = \frac{Y'}{Y} = \frac{n_0 \sin U_0}{n_k \sin U_k}. \tag{2.30}$$

where Y and Y' are the object and image heights. If a lens system is to be corrected for spherical and coma aberrations, they must have the same magnification for the marginal rays (M) and for the paraxial rays (m). Then, the following expression, called the *optical sine condition*, may be obtained as follows

$$\frac{\sin U_0}{u_0} = \frac{\sin U_k}{u_k}. \tag{2.31}$$

The product $n_k \sin U_k$ is called the numerical aperture (NA) and determines the diameter of the diffraction image point. The NA is often given in the specification of a lens system. When the system works at finite conjugates, this NA value also specifies the $n_0 \sin U_0$ on the object side, from Eq. (2.30). Since it is customary to have the paraxial rays pass through the entrance pupil at the same height as the extreme marginal ray, the value of the paraxial u_0 is taken at $\tan U_0$.

When the object point is at infinity, Eq. (2.30) reduces to

$$Y_1 = \frac{-f'}{\sin U_k}. \qquad (2.32)$$

In this case, the second principal refracting surface of the lens is a sphere passing through P_2, the second principal point, and with its center on the paraxial focus.

When the paraxial ray is traced through the lens at a height of $y_1 = Y_1$, the final paraxial angle u_k will be less than U_k, because $u_k = \tan^{-1}(y_1/f')$ rather than $\sin^{-1}(y_1/f')$.

The terms $F/$number or $(F/\#)$ are frequently used instead of the numerical aperture. The definition of the $F/$number is

$$F/\text{number} = \frac{f'}{D}, \qquad (2.33)$$

where D refers to the diameter of the axial beam on the entrance pupil when the object is at infinity. In this case:

$$F/\text{number} = \frac{f'}{2Y_1} = \left|\frac{1}{2\sin U_k}\right| = \frac{n_k}{2\text{NA}_k}. \qquad (2.34)$$

When the lens operates at finite conjugates, then the effective $F/$number on the image side is given by

$$(F/\text{number})_{\text{eff}} = |(1-m)|(F/\text{number})_{\text{inf}}. \qquad (2.35)$$

2.3.2. The Cardinal Points of a Lens

By tracing two paraxial rays, the six cardinal points for a lens may be found. This subject is treated in every textbook on optics. It is presented here for completeness.

Figure 19 shows the six cardinal points of a lens when the object index of refraction is n_0 and the image side index of refraction is n_k. The points F_1, F_2, P_1, P_2 are located by tracing a paraxial ray parallel to the optical axis at a height of unity. A second ray is traced parallel to the optical axis but from right to left. The incoming rays focus at F_1 and F_2 and appear to refract at the planes P_1 and P_2. This is why they are called *principal planes.* These planes are also planes of unit magnification. The nodal points are located by tracing a ray back from C parallel to the ray direction F_1A. If this later ray is extended back towards the F_2 plane, it intersects the optical axis at N_1. This construction to locate N_1 and N_2 shows that a ray which

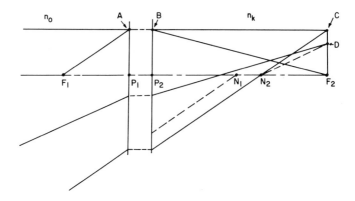

FIG. 19. The location of the cardinal points in a lens system.

enters the lens headed towards N_1 emerges on the image side from N_2 making the same angle with the optical axis.

In the diagram, a third ray is traced through the lens, and it intersects the image plane at the point D.

The following relations hold for any lens:

$$F_1 P_1 = f \quad \text{and} \quad P_2 F_2 = f', \tag{2.36}$$

$$N_1 N_2 = P_1 P_2, \tag{2.37}$$

$$F_1 P_1 = N_2 F_2, \tag{2.38}$$

$$F_1 N_1 = P_2 F_2, \tag{2.39}$$

$$\frac{f}{n} = \frac{f'}{n'}. \tag{2.40}$$

When the object distance is infinite, it is customary to trace the a and b rays as explained in Section 2.2.6. The tracing of these two rays provides enough information to locate all the cardinal points. An excellent exercise is to graphically locate the cardinal points when all of the information one has is the ray diagram shown in Fig. 20.

When the object plane is at a finite distance, the a and b rays are not traced in parallel to the optical axis. Then the two rays do not directly locate the focal points and the principal planes. It is however possible to locate the cardinal planes from the two rays traced from the finite object by using Eqs. (2.25) and (2.26). The A and B constants can be evaluated, and then the paths of the two rays entering parallel to the optical axis can

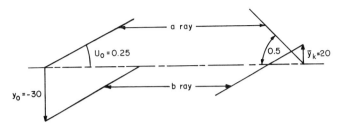

FIG. 20. A diagram showing the *a* and *b* rays entering and exiting an optical system. One can locate all the cardinal points from this diagram.

be determined. The following equations can thus be derived for the cardinal point locations.

$$f' = \frac{H}{(u_k \bar{u}_0 - u_0 \bar{u}_k) n_0}, \tag{2.41}$$

$$f = \frac{f' n_0}{n_k}, \tag{2.42}$$

$$V_2 F_2 = \frac{(\bar{y}_{k-1} u_0 - y_{k-1} \bar{u}_0) n_0 f'}{H}, \tag{2.43}$$

$$F_1 V_1 = \frac{(\bar{y}_1 u_{k-1} - y_1 \bar{u}_{k-1}) n_0 f'}{H}. \tag{2.44}$$

When the cardinal points of a lens have been located, the image of any object can be found by using the following relations, as illustrated in Fig. 21.

$$m = \frac{y_k}{y_0} = \frac{-x'}{f'} = \frac{-f}{x}, \tag{2.45}$$

$$xx' = ff', \tag{2.46}$$

$$\frac{f}{s} + \frac{f'}{s'} = 1. \tag{2.47}$$

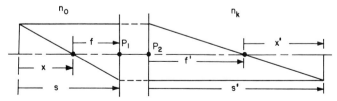

FIG. 21. Conjugate relations for object and image distances.

When the lens is surrounded by air, these equations reduce to

$$m = -\frac{x'}{f'} = -\frac{f'}{x}, \tag{2.48}$$

$$xx' = f'^2, \tag{2.49}$$

$$\frac{1}{s} + \frac{1}{s'} = \frac{1}{f'}. \tag{2.50}$$

The second expression is the Newton relation. One of the easiest ways to locate the cardinal points in the laboratory is to image a distant source at F_1 and measure the distance to some reference point on the lens. Then turn the lens around and image the distant object on the point F_2 and measure from the same reference point on the lens. Place an object a distance x from F_1 and measure the distance x' from F_2. Solve for f' and use it to measure from F_1 and F_2 to locate P_1 and P_2.

2.3.3. Thin Lenses

In planning lens systems, it is convenient to use the concept of thin lenses. The thickness of the lens is considered to be zero. The refraction equation for the thin lens is

$$u' = u + y(-\phi), \tag{2.51}$$

where

$$\phi = (n-1)(c_1 - c_2).$$

Table II shows the computation sheet for analyzing a system of three thin lenses.

TABLE II. Calculation Sheet for Thin Lens, which Includes Thin-Lens Chromatic Aberration Calculations

$-\phi$		-0.1510	0.3097	-0.2006		
t			1.620	1.067	8.252	
y		1.43	1.080	1.180	0	$f' = 10$
μ	0		-0.216	0.0937	-0.1430	
$\bar{y}$		-1.0692	0	0.7042	4.986	
$\bar{\mu}$	0.4986		0.66	0.66	0.4594	
V		60.23	36.70	60.23		
$y^2\phi/V_{U_R}$		-0.0357	$+0.0685$	-0.0322	Tach $= 0.0007$	
$y\bar{y}\phi/V_{U_R}$		-0.0266	0	0.0189	Tch $= +0.0077$	

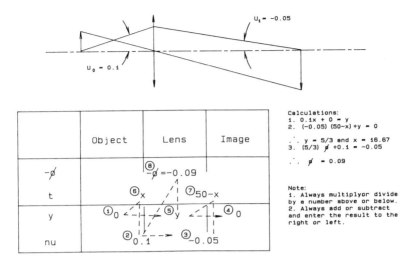

FIG. 22. The y nu method to solve imaging problem 1.

2.3.4. Sample Calculation Sheets for Thin Lenses

The paraxial calculation of optical systems will be illustrated by means of three examples.

(a) Problem 1. Design an optical system to relay an image with a magnification m equal to -2, a numerical aperture NA on the image side equal to -0.05, and an object-to-image distance equal to 50 cm.

A value of NA equal to 0.1 means that $\sin U_k = 0.1$ and U_k is equal to 5.74. Then, $\tan(5.74) = 0.1005$; therefore, we use $u_0 = 0.1$. Now, $n_k u_k$ may be calculated by using Eq. (2.29), obtaining a value equal to -0.05.

A sketch of the thin lens with its calculations is illustrated in Fig. 22.

(b) Problem 2. Redesign the system specified in Problem 1, but change the magnification to $+2$ in order to produce an erect image.

The diagram in Fig. 23 shows that two lenses are now required. It is easy to see that there is an infinite number of solutions for the positions of the lenses. The first lens can work at equal conjugates, and both lenses can have equal power ($\phi_1 = \phi_2$) (focal lengths).

The solution may be obtained in the following six steps:

1. Pick $y_1 = 1$.
2. Set $u_2 = -0.1$ because lens 1 works at equal conjugates.
3. ϕ_1 and ϕ_2 must both be equal to 0.02.
4. Solve for $y = -0.15/0.2 = -0.75$.

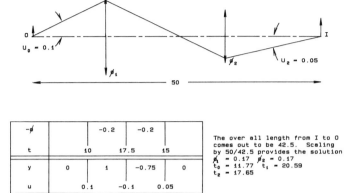

FIG. 23. The $y\,nu$ method to solve imaging problem 2.

5. Solve for t_0, t_1, t_2.
6. The object-to-image distance is then 42.5.

The system can then be scaled by the ratio of 50/42.5. The calculation sheet is shown in Fig. 23. Lens 1 works at $F/2.5$ and should probably be a triplet if it is to be color-corrected. Lens 2 works at $F/3.3$, and a doublet achromat should be sufficient. Lens 1 could be made out of two doublet objectives of the same focal length and with parallel light between them. This would be more elements, but there is a much better chance that "on-the-shelf" lenses could be found.

(c) Problem 3. Design a telescope with a magnifying power MP = 10×, whose objective diameter is 50 mm and whose F/number is 5. The real half field of view is 3.5° and the eye relief is 18 mm.

The system has an objective, a field lens, and an eye lens, as shown in Fig. 24. The calculation sheet for this lens system is shown in the same figure. Before the powers of lenses 2 and 3 are known, the a and b rays can be traced through the system. The powers may be solved for when the ray data is complete.

2.3.5. Chromatic Aberrations

The paraxial-ray tracing and the subsequent location of the cardinal points in a lens depend upon the index of refraction of the glass. The index is wavelength-dependent and should be written as n_λ. An achromatic lens is a lens that is corrected so that the cardinal points are located in the same positions for a band of wavelengths. Therefore, the lens has to focus a band of wavelengths at the same focal position and have the same focal length.

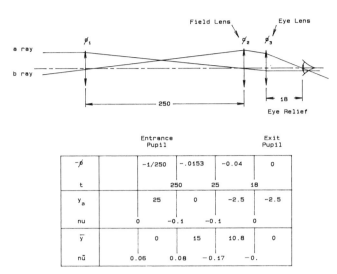

		Entrance Pupil			Exit Pupil
$-\phi$		$-1/250$	$-.0153$	-0.04	0
t			250	25	18
y_a		25	0	-2.5	-2.5
nu	0		-0.1	-0.1	0
$\bar{y}$		0	15	10.8	0
$n\bar{u}$	0.06		0.08	-0.17	$-0.$

FIG. 24. The $y\,nu$ method to solve imaging problem 3.

To realize this, the points F_1, F_2, P_1, P_2 must be the same for λ and $\lambda + \Delta\lambda$. Figure 25 shows an achromatic lens made up of three spaced lenses. The positive lenses have low dispersion, and the negative lens has high dispersion. It is clear that the axial ray and the oblique ray must separate for the different wavelengths upon entering the lens. To be achromatic, it is necessary for the color rays to emerge from the same position on the rear surface. The spreading of the rays in the diagram has been exaggerated in order to see the separation, but they are in the correct relative positions.

By differentiating Eq. (2.16), it is possible to derive an equation for the transverse axial chromatic (Tach) aberration that each surface contributes to the final image. The surface contribution for the j surface is

$$\text{Tach } C_{\lambda_2-\lambda_3} = yni\frac{\left(\dfrac{dn}{n} - \dfrac{dn'}{n'}\right)}{n_k u_k}, \qquad (2.52)$$

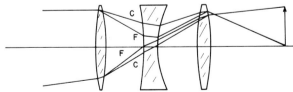

FIG. 25. A diagram illustrating ray paths for F and C light when the lens has been corrected for Tach and Tch.

y is the height of the axial ray traced at the wavelength λ_1.
n is the index of refraction at the wavelength λ_1.
i is the angle of incidence on the surface for the wavelength λ_1,
where

$$i = yc + u. \tag{2.53}$$

dn is the difference in index of refraction between λ_2 and λ_3. λ_1 is a wavelength approximately midway between λ_2 and λ_3.

Tach $(\lambda_2 - \lambda_3)$ is $(y_k, \lambda_2 - y_k, \lambda_3)$. It is the ray displacement between λ_2 and λ_3 as measured in the image plane λ_1. When Tach $(\lambda_2 - \lambda_3)$ is positive, the λ_3 ray intersects the image plane above the λ_2 ray, and the lens is said to be overcorrected for color.

Tach C_j is the contribution of the j surface to the total system Tach, and

$$\text{Tach} = \sum_{j=1}^{k} \text{Tach } C. \tag{2.54}$$

The Tach aberration of a lens is illustrated in Fig. 26.

The oblique chief ray contributes to the chromatic aberration of the image height. The formula is the same as for the axial Eq. (2.54), except for the angle $\bar{i}$, which refers to the oblique ray. The equation is

$$\text{Tch } C = \left(\frac{\bar{i}}{i}\right) \text{Tach } C. \tag{2.55}$$

The total Tch is also the summation of all the surface contributions. Table III is a calculation sheet of the chromatic calculations of the axial and oblique rays for an air-spaced triplet.

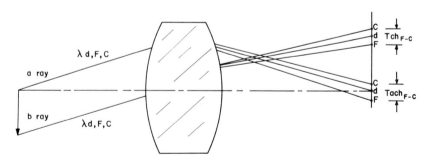

FIG. 26. The chromatic aberrations of the a and b rays.

GAUSSIAN OPTICS 33

TABLE III. Calculation of Tach and Tch for a Triplet Objective

c		0.2339		−0.0098		−0.2107		0.2516		0.06799		−0.2530		Image Plane
t			0.6		1.211		0.25		1.016		0.6		8.33603	
n	1		1.62041		1		1.61659		1		1.62041			
$(n-n')c$		−0.14511		−0.00608		0.12992		0.15513		−0.04218		−0.15696			$f' = 1$
t/n	∞		0.37028		1.211		0.15465		1.016		0.37028		8.33603	
y_a	1.43		1.35316		1.09190		1.08048		1.17571		1.19205		0	
nu_a	0		−0.20751		−0.21574		−0.07389		0.09373		0.04414		−0.14297	
y_b		−1.16769		−0.91981		−0.10233		0		0.67234		0.90686		0.5
nu_b	0.5		0.66944		0.67504		0.66175		0.66175		0.63339		0.49102	
dn/n	0		0.00635		0		0.01042		0		0.00635		0	
dn'/n'		−0.00635		0.00635		−0.01042		0.01042		−0.00635		0.00635		
Tach C		−0.02126		−0.01377		0.03550		0.02877		−0.00908		−0.02356		Tach = −0.00337
Tch C		−0.01442		0.04116		−0.05544		0.05215		−0.03700		0.01386		Tch = 0.00031

2.3.6. Thin-Lens Chromatic Aberrations

Further simplification can be achieved by combining the two surfaces of a lens to obtain the chromatic-aberration contributions for a thin lens. The thin-lens contributions to the two chromatic aberrations are given by the equations

$$\text{Tach } C = \frac{1}{(nu)_k} \cdot \frac{y^2}{f} \cdot \frac{dn}{n_d - 1}, \tag{2.56}$$

$$\text{Tch } C = \frac{\bar{y}}{y} \text{Tach } C. \tag{2.57}$$

When two thin lenses (A and B) are separated with zero spacing, an achromat can be formed where the following holds.

$$\frac{\left(\frac{dn}{n-1}\right)_A}{f_A} + \frac{\left(\frac{dn}{n-1}\right)_B}{f_B} = 0. \tag{2.58}$$

The ratio $dn/(n-1)$ is the ratio of dispersion to ray bending.

To assure an accurate value of the index of refraction of the glass, the index is measured at spectral lines. Most visual achromats are computed by using $dn = n_F - n_C$; and n_d is used as the intermediate index. The F light λ is 485.6 nm, C light is 656.28 nm, and d light is 587.56 nm. F and C light are hydrogen lines, and d light is a helium line. Glasses are referred to by code numbers using three digits for the V number. The V number is equal to $(n_d - 1)/(n_F - n_C)$. For example, the code number 517635 refers to the glass with $n_d = 1.517$ and $V = 63.5$.

Figure 27 shows a plot of the transverse axial color for a visual achromat as a function of the wavelength. Note the following about the curve.

(a) Only two wavelengths at a time are brought to a common focus.

(b) The minimum change of focus for a change in wavelength occurs at $\lambda = 0.55$ μm, which is the peak of the sensitivity of the human eye. This wavelength is not quite midway between the F and C light.

(c) There is transverse chromatic aberration between the common F and C focus and the d focus. This is called *secondary color* or *residual color*.

In a thin-lens doublet, this is equal to

$$\text{Tach}_{F-d} = \frac{D(P_A - P_B)}{2(V_A - V_B)}, \tag{2.59}$$

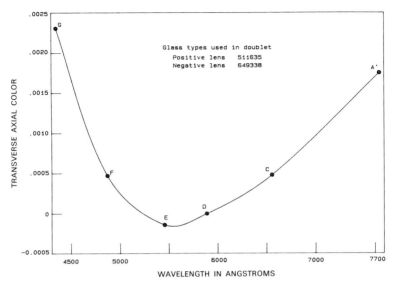

FIG. 27. The transverse axial chromatic aberration of a lens as a function of wavelength.

where $P = (n_F - n_C)/(n_F - n_d)$, $V = (n_d - 1)/(n_F - n_C)$, and D is the clear aperture diameter of the lens.

Figure 28 is a plot of P versus V for a large selection of glasses. As one can see, most of the achromats will have a constant value of $(P_A - P_B)/$

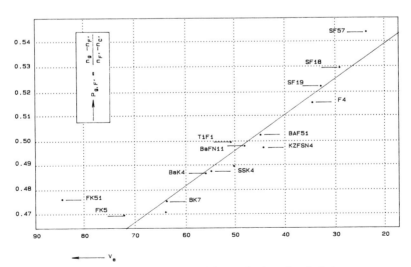

FIG. 28. A plot of P versus V for a selection of typical glasses.

($V_A - V_B$). The slope of the line shown on the graph is 1/1600, so the two glasses lying on this line will have

$$\text{Tach}_{F-d} = \frac{D}{3200}.$$

There are a few glasses which fall off the line far enough to reduce the secondary color. The most outstanding examples are 613442, flourite, and quartz.

These materials should be used with caution, for the following reasons. Glass 613442 is expensive and tends to tarnish and stain easily. Flourite is expensive and comes in small pieces (less than one inch in diameter). It is used primarily in microscope objectives. It does not polish as smoothly as glass and introduces scattered light. If used, it should be cemented between glass elements to reduce the scattering. The phosphate glasses are also expensive and tarnish easily.

The most effective way to reduce the secondary color in optical systems is to keep the F/number of the lenses large. When small F/numbers are essential, the focal lengths should be kept short. These rules are illustrated in the following two cases.

(1) A terrestrial telescope may use large F/number objectives by making the focal lengths of the objective and the eyepiece long. The secondary color comes primarily from the objective because its aperture is large. The objective secondary blur is equal to $D/3200$. By making the focal length of the eyepiece large, the angular subtense of the secondary color is reduced. The result is that the telescope should be made as long as possible.

(2) In a microscope, one is forced to use a small F/number objective in order to obtain high resolution. For this reason, the microscope objectives have focal lengths ranging from around 2 mm to 32 mm. In the short focal lengths, the working distance becomes extremely small. The working distance can be increased, but at the expense of increasing the secondary color.

In dealing with secondary color in instruments, it is often a mistake to add complexity in order to reduce the secondary color below a detectable amount. The eye is remarkably capable of visual acuity with secondary color present. This is particulary true for instruments used for pointing. It is more important to form images which are symmetrically round. By adding extra elements with stronger curves, the lens may be manufactured with decentering, which would destroy any gain from secondary color correction.

C. G. Wynne[6,7] has pointed out that Eqs. (2.52) and (2.55) are not strictly true. The chromatic blurs are not additive when the chromatic blurs become large. This error does not cause difficulty in most practical cases, but it has

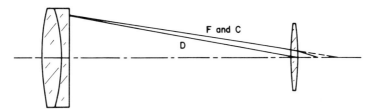

FIG. 29. An example of how the secondary color can be reduced by placing a lens in a position where there is a substantial spreading of the color rays.

masked thinking of possible ways to correct secondary color, because Eq. (2.59) suggests that secondary color cannot be corrected with ordinary glasses using a large V difference.

Figure 29 is an example showing why the simple theory is not correct. The figure shows an exaggerated drawing of an achromat with secondary color. A lens placed in the focal plane of the d light will have no effect on the d light, however, the F and C light rays will be bent and will reduce the secondary color, adding to the Petzval sum which is field curvature. (See Section 2.6 for the Petzval sum). This principle is used in microscope objectives. They do not have as much secondary color as their focal lengths suggest, but they have more field curvature. When flat-field microscopes are developed, it is important to expect more secondary color. The decision to use special glasses then is sometimes necessary, which explains its higher cost.

2.4. Stops, Pupils, Vignetting, and Telecentricity

Most optical systems should have an aperture stop to control the cone angle of flux entering the lens from the object. The diaphragm of a camera or the iris of the human eye are examples of aperture stops. Usually, but not always, the aperture stop is surrounded by lenses, as shown in the simple system illustrated in Fig. 30. The following should be noted about the diagram.

1. The aperture stop limits the angle u_0 and u_k.
2. The chief ray passes through the center of the aperture stop.
3. The chief ray is directed towards the center of the entrance pupil and emerges from the center of the exit pupil.
4. The field stop limits the angle of obliquity for the chief ray.

The entrance and exit pupil are usually located by tracing the axial a and the chief b ray.

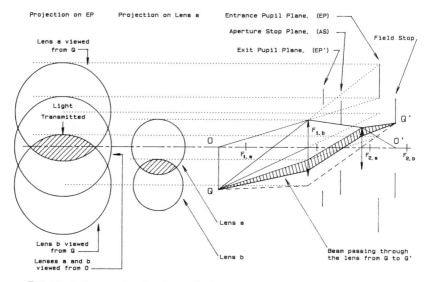

FIG. 30. A diagram showing the pupil planes and field stop for a two-lens system.

Figure 30 shows the diameters of the a and b lens and the aperture stop as they project onto the entrance pupil, as viewed along the chief ray. The common area is the shape of the beam that passes through the lens for the off-axis point Q. The lenses a and b vignette the full aperture. This causes a reduction in the flux passing through the lens, and it influences the shape of the image of the point. Most photographic lenses vignette as much as 50% of the flux that can pass through the aperture stop.

The field stop in Fig. 30 is usually the plate or film holder in a camera. The field stop should be located in an image plane in order to have a sharp cutoff in the field. If the film plane had no mechanical limit to the film area, the image brightness would fall off gradually.

Optical systems are sometimes designed to image the exit pupil at infinity. The chief ray then emerges parallel to the optical axis; and if there is no vignetting, the perfectly corrected off-axis image is exactly the same as the central image. This avoids an obliquity factor on the receiver and means that slight focus variations do not result in a change of image size. Lenses of this type are called telecentric on the image side. Lenses can also be telecentric on the object side.

The size, shape, and position of the exit pupil has a strong effect on the shape of the image of a point source if the lens is near diffraction-limited. Too much vignetting can cause an image to appear astigmatic. The exit subtend of the exit pupil also determines the illuminance incident on the image plane.

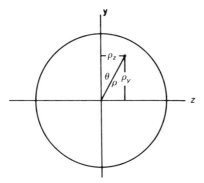

FIG. 31. The coordinates of a general ray passing through the paraxial entrance pupil.

The paraxial entrance pupil is used as a reference plane for all the rays traced through the optical system. Figure 31 shows the coordinate system in the entrance pupil of a lens as it is seen from the object plane. The paraxial a ray passes through at the height y_1, and the b ray passes through the center of the pupil. The coordinates of a general ray are Y_1, Z_1. Usually these coordinates are specified relative to the paraxial height y_1 as

$$\rho_y = \frac{Y_1}{y_1}, \tag{2.60}$$

$$\rho_z = \frac{Z_1}{y_1}, \tag{2.61}$$

where

$$\rho_y = \rho \cos \theta \quad \text{and} \quad \rho_z = \rho \sin \theta. \tag{2.62}$$

The value of ρ normally ranges from zero to 1 and is called the *fractional height* (FY).

There usually is some aberration of the pupil planes. A real ray traced through the center of the paraxial entrance pupil may not pass through the center of the aperture stop. In spite of this, the paraxial pupil plane is usually used as the reference surface for the pupils, and for most lenses the pupil aberrations are small. In wide-angle systems, however, one must consider the effect of pupil movements. It is pupil aberration that makes it difficult to view through a wide-angle telescope. It is necessary to have an oversize pupil diameter to provide oblique rays that can enter the eye pupil.

2.5. Paraxial-Ray Summary

It may appear odd that optical engineers are so concerned with paraxial rays, when they represent an infinitesimal group of rays surrounding the optical axis. They are important for the following reasons.

(1) If the lens is to be aberration-free, the rays of finite aperture must come to the same image positions as the paraxial rays. Any deviation from the paraxial-ray images represents an image error.

(2) It is possible, by using paraxial ray data on each surface of the lens, to calculate the first-order chromatic and the third-order monochromatic aberrations. The surface contributions to the image errors indicate the source of the aberrations.

(3) In a well-corrected lens system, the paraxial construction lines and the actual rays deviate in path from each other by small amounts. The deviations are usually less than the tolerance needed to determine the clear apertures of the lenses.

(4) In this day of large computers, a designer spends much of the time in setting up lenses with the proper paraxial configurations. If any hand computing is done, it is concerned with paraxial rays. The ray tracing and image-error reduction is done entirely on the large computer. It is through thoroughly understanding first-order paraxial optics and the sources of third-order aberration that a designer knows how to specify starting points for the automatic correction of image errors.

2.6. Third-Order Aberrations. The Aberration Polynomials

The time has come to get off the optical axis and note what happens when rays are traced through the lens at finite heights above the optical axis. Figure 32 shows the object plane, the entrance plane, and the image plane as viewed from the object surface and looking towards the image plane.

2.6.1. Third-Order Aberrations

In a system which is rotationally symmetric around the optical axis, it is completely general to consider object points along a single radial vector from the optical axis. All object points at a given radius from the optical axis are imaged by the lens in exactly the same way. An object point (Y_0) can be specified relative to the height of the b paraxial ray on the object plane as follows: FOB = Y_0/y_0. FOB is the fractional object height. The general ray passes through the entrance pupil at the coordinates Y_1, Z_1, and it intersects the image plane at Y_k, Z_k. The entrance-pupil coordinates of the general ray are specified in coordinates relative to the paraxial ray,

$$\rho_y = \frac{Y_1}{y_1}, \qquad (2.63)$$

$$\rho_z = \frac{Z_1}{y_1}. \qquad (2.64)$$

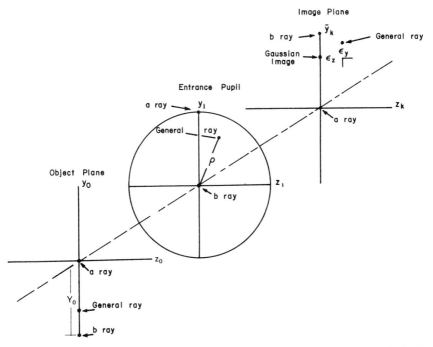

FIG. 32. View of the object plane; entrance pupil as seen looking along the optical axis from the object side and towards the image plane.

When the optical system is aberration-free, the general ray should intersect the image plane at $Y_k = \bar{y}_0 \cdot m \cdot \text{FOB}$ and $Z_k = 0$. Displacements from this are measures of an image error given by

$$\varepsilon_y = (Y_k - \bar{y}_0 \cdot m \cdot \text{FOB}), \tag{2.65}$$

$$\varepsilon_z = Z_k. \tag{2.66}$$

The ray displacements in ε_y and ε_z can be expressed as a series of displacements of various orders,

$$\varepsilon_y = \varepsilon_{3y} + \varepsilon_{5y} + \cdots, \tag{2.67}$$

$$\varepsilon_z = \varepsilon_{3z} + \varepsilon_{5z} + \cdots. \tag{2.68}$$

The third-order terms written in the relative coordinates are

$$\varepsilon_{3y} = \sigma_1 \rho^3 \cos\theta + \sigma_2 \rho^2 (2 + \cos 2\theta)(\text{FOB})$$
$$+ (3\sigma_3 + \sigma_4)\rho \cos\theta (\text{FOB})^2 + \sigma_5 (\text{FOB})^3 \tag{2.69}$$

$$\varepsilon_{3z} = \sigma_1 \rho^3 \sin\theta + \sigma_2 \rho^2 \sin 2\theta (\text{FOB}) + (\sigma_3 + \sigma_4)\rho \sin\theta (\text{FOB})^2 \tag{2.70}$$

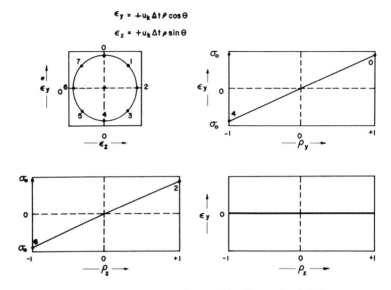

FIG. 33. Transverse ray plots resulting from a focal shift.

The coefficients are

1. σ_1 = third-order spherical aberration
2. σ_2 = sagittal coma
3. σ_3 = sagittal astigmatism
4. σ_4 = Petzval coefficient
5. σ_5 = distortion

When the image plane is shifted by an amount Δt_k, focal shift terms must be added to the above polynomial. They are

$$\varepsilon_y = u_k \cdot \Delta t_k \cdot \rho \cos \theta, \qquad (2.71)$$

$$\varepsilon_z = u_k \cdot \Delta t_k \cdot \rho \sin \theta. \qquad (2.72)$$

The interpretation of these coefficients are shown in Figs. 33 to 36. The distortion term represents a displacement of the Y position of the chief ray.

2.6.2. The Wave-Aberration Polynomial

To evaluate a well-corrected optical system, it is necessary to consider the optical paths (OP) from the object point to the image point. In a perfectly corrected system, the optical path from the object to the image point is a

THIRD-ORDER ABERRATIONS. THE ABERRATION POLYNOMIALS 43

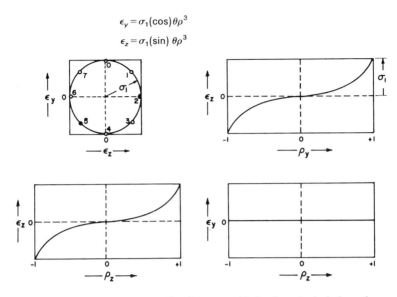

FIG. 34. Transverse ray plots resulting from pure third-order spherical aberration.

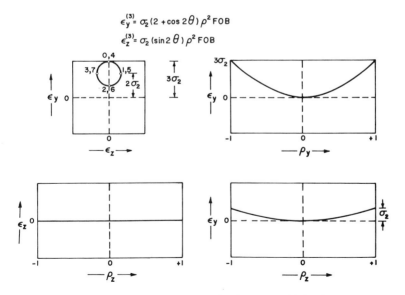

FIG. 35. Transverse ray plots resulting from pure third-order coma.

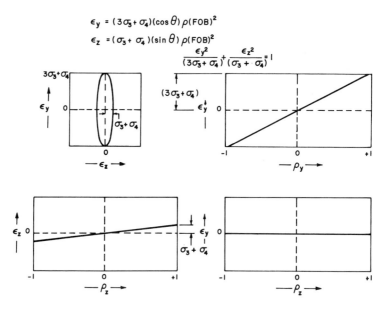

FIG. 36. Transverse ray plots resulting from pure third-order astigmatism and field curvature.

constant for all ray paths traced through the entrance pupil. This means that the wave front converging to the image point is a spherical surface. When there is aberration, the converging wave front is not spherical. In order to specify the path length for an aberrated ray, the path length is measured along the ray from the object point to a line drawn through the image point and normal to the ray. This is illustrated in Fig. 37. This measure of the path length is justified by the concept that the emerging aberrated wave front can be considered as the superposition of an infinite number of plane waves which are normal to the rays. The difference between the perfect reference sphere and the aberrated wave front (OPD) may also be expressed by an aberration polynomial. The OPD polynomial for the Y, X plane is

$$\text{OPD} = a_1\rho^2 + a_2\rho(\text{FOB}) + b_1\rho^4 + b_2\rho^3(\text{FOB})$$
$$+ 3b_3\rho^2(\text{FOB})^2 + b_4\rho^2(\text{FOB})^4 + b_5\rho(\text{FOB})^3. \quad (2.73)$$

The term a_1 represents a focal shift from the paraxial focal plane. The term a_2 is the coefficient for a transverse shift of the center of the reference sphere from the Gaussian image plane.

THIRD-ORDER ABERRATIONS. THE ABERRATION POLYNOMIALS

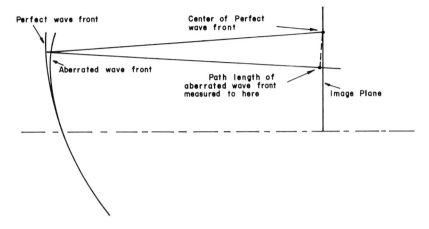

FIG. 37. The optical path of an aberrated ray is measured to the ray normal which passes through the perfect image point.

The wave-front errors and the ray displacements are related through the expressions:

$$d\,\text{OPD}/dY_1 = \frac{-\varepsilon_y n_k}{f'}, \tag{2.74}$$

$$d\,\text{OPD}/dZ_1 = \frac{-\varepsilon_z n_k}{f'}. \tag{2.75}$$

By differentiating the expression for OPD in Eq. (2.73), the connection between the wave-front coefficients and the transverse ray displacements may be computed from the following equations.

$$b_1 = \frac{-\sigma_1 n_k}{8F/\#}, \tag{2.76}$$

$$b_2 = \frac{-\sigma_2 n_k}{2F/\#},$$

$$b_3 = \frac{-\sigma_3 n_k}{4F/\#}, \tag{2.78}$$

$$b_4 = \frac{-\sigma_4 n_k}{4F/\#}, \tag{2.79}$$

$$b_5 = \frac{-\sigma_5 n_k}{2F/\#}. \tag{2.80}$$

Some designers think in terms of OPD wave-front deviations expressed in wavelengths. This is appropriate when designing optical systems with near diffraction-limited images. There are designs, however, where the OPD values are many wavelengths, and in that case, the size of the image can be computed from the geometrical deviations. For these problems, it is perhaps more convenient to think of transverse image errors. Prior to large computers, it was difficult to compute accurate optical path lengths. As a result, designers concentrated on ray deviations in the image plane, and the OPD values were calculated by integrating Eqs. (2.74) and (2.75). Now that large computers are in use, direct calculation of the optical path is commonly used.

2.6.3. The Calculation and Use of Third-Order Aberration Coefficients

The third-order aberration coefficients are used extensively in optical-system planning and design.[8] They are important for the following reasons.

(1) The coefficients can be calculated entirely from the paraxial data.

(2) The coefficients are the summation of individual surface contributions. They provide insight as to the source of aberrations and lead to understanding of how to correct image errors and even how to balance out higher-order image errors.

(3) The third-order aberrations must be well corrected in order to obtain a well-corrected lens system.

(4) The third-order theory is simple enough to enable one to understand how to correct a lens. It provides some assurance that the design is located in a configuration that has a chance of success.

(5) Thin-lens equations for the third-order aberrations are simple and can be calculated easily.

The third-order coefficients may be computed from the paraxial data by using the following formulae.

$$\sigma_1 = \frac{1}{2(nu)_k} \sum_{j=1}^{k} (Si^2)_j, \qquad (2.81)$$

$$\sigma_2 = \frac{1}{2(nu)_k} \sum_{j=1}^{k} (Si\bar{i})_j, \qquad (2.82)$$

$$\sigma_3 = \frac{1}{2(nu)_k} \sum_{j=1}^{k} (S\bar{i}^2)_j, \qquad (2.83)$$

$$\sigma_4 = \frac{1}{2(nu)_k} \sum_{j=1}^{k} P_j, \qquad (2.84)$$

THIRD-ORDER ABERRATIONS. THE ABERRATION POLYNOMIALS

$$\sigma_5 = \frac{1}{2(nu)_k} \sum_{j=1}^{k} S i \bar{i} + H(\bar{u}'^2 - \bar{u}^2), \tag{2.85}$$

where

$$S = -yn\left(\frac{n}{n'-1}\right)(i+u'), \tag{2.86}$$

$$i = yc + u, \tag{2.87}$$

$$P = c\left(\frac{n-n'}{nn'}\right), \tag{2.88}$$

$$\bar{S} = -yn\left(\frac{n}{n'-1}\right)(\bar{i}+\bar{u}'), \tag{2.89}$$

$$H = \bar{y}_0 n_0 u_0. \tag{2.90}$$

These equations may be applied to a thin lens with the chief ray passing through its center and yield the following equations for the aberrations of the thin lens.

$$\sigma_1 = \alpha_1 + \alpha_2 c_1 + \alpha_3 c_1^2 \tag{2.91}$$

$$\sigma_2 = \beta_1 + \beta_2 c_1, \tag{2.92}$$

$$\sigma_3 = \frac{H^2}{2f'(nu)_k}, \tag{2.93}$$

$$\sigma_4 = \frac{-H^2 P}{2(nu)_k}, \tag{2.94}$$

$$\sigma_5 = 0. \tag{2.95}$$

When the chief ray passes through the lens away from the center of the lens, the coefficients may be computed from the stop-shift equations:

$$\text{Tach}^* = \text{Tach}, \tag{2.96}$$

$$\text{Tch}^* = \text{Tach} \cdot Q + \text{Tch}, \tag{2.97}$$

$$\sigma_1^* = \sigma_1, \tag{2.98}$$

$$\sigma_2^* = Q\sigma_1 + \sigma_2, \tag{2.99}$$

$$\sigma_3^* = Q^2 \sigma_1 + 2Q\sigma_2 + \sigma_3, \tag{2.100}$$

$$\sigma_4^* = \sigma_4, \tag{2.101}$$

$$\sigma_5^* = Q^3 \sigma_1 + 3Q^2 \sigma_2 + Q(3\sigma_3 + \sigma_4) + \sigma_5. \tag{2.102}$$

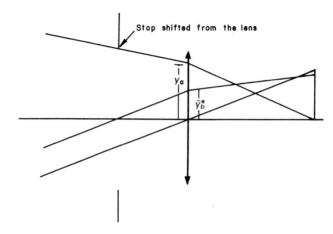

FIG. 38. A diagram showing a stop shift and the data used to compute Q.

The quantities with * are the new coefficients obtained by having the stop removed from the lens. Q is $\bar{y}^*/y$, where $\bar{y}^*$ is the height of the new chief ray on the lens, and y is the height of the axial ray. This is illustrated in Fig. 38.

The following example shows how one can devise concepts to shape lenses and correct aberrations. The lens is a single plano-convex lens with a focal length of 10 mm and an F-number of 10. It illustrates how Eqs. (2.98) and (2.102) are used qualitatively as a guide to a solution.

A single lens shaped to minimize spherical aberration, with the stop in contact with it, will have a meridional ray plot for a 20° off-axis point, as shown in Fig. 39. The curve is essentially a straight line with a negative slope. This focus is closer to the lens than to the paraxial image plane. The image error at $\rho = 1$ is -0.13. The focus will then be at -2.6 units ($-0.13/0.05$) from the paraxial image plane. Plotted in the same figure are the terms due to σ_3 and σ_4. Equation (2.69) shows that at $\rho = 0$, the slope of the line of ε_y versus ρ is $(3\sigma_3 + \sigma_4)$, which is astigmatism and Petzval field curvature.

Equations (2.93) and (2.94) show that σ_3 and σ_4 will not change as the lens is shaped into another form as long as the stop is at the center of the lens.

In order to reduce the slope of the meridional ray plot, it is necessary to remove the stop from the lens. Equation (2.100) shows that σ_3^* will be influenced by the values of σ_1 and σ_2. The sigma σ_1 of a single lens will always be negative, so the term $Q^2\sigma_1$ will be added to the negative σ_3 and make the slope more negative. However, if the lens is shaped by bending sharply to the left, σ_2 will be positive. This can be understood by observing

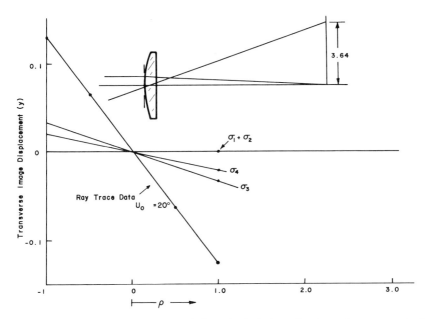

FIG. 39. A single lens shaped to minimize spherical aberration has a large amount of inward-curving field.

the bent lens shown in Fig. 40. The upper ray will pass through the lens at close to minimum deviation. The lower ray is far from minimum deviation, so that it will be refracted more than the upper ray, and the two rays will intersect above the central ray. This is positive coma. The positive coma with a positive Q will add plus σ_3^*, and it may be possible to have it more than equalize σ_3 and $Q^2\sigma_1$, thereby providing a flat meridional fan. Figure 40 shows the ray tracing of the lens bent far to the left. The ray plot is essentially a cubic curve. By selecting the rays surrounding the inflection point, the field is considerably flatter than in Fig. 39. This has been achieved, however, by introducing more σ_1; so the lens cannot be used at a large aperture.

There are other things to note about the curve. The inflection point for the new chief-ray position is below the image height for the old chief ray. Equation (2.95) shows that the distortion is zero for the lens with the stop in contact. When shifting to the new stop, the chief ray will intersect the image plane at 0.08 units below the Lagrange image height. This is minus 2% distortion.

A close look at the curve around the inflection point shows that the curve is not completely symmetrical. The ray at $\rho = 3$ is dropping too fast. This is due to higher-order spherical aberration. The lens has been bent into a

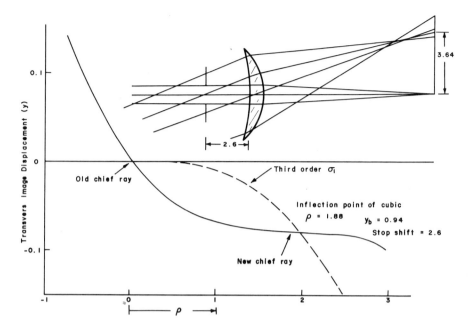

FIG. 40. A single lens bent to introduce positive coma. When the stop is moved to the left, a position for a flat field can be found.

shape which introduced large third-order spherical aberration. The third-order spherical aberration of the lens with stop removed is shown as a dotted line in Fig. 40. The upper ray for the 20° beam is incident at a value of $\rho = -2.8$. At this value of ρ, the third-order spherical is -0.24. Some high-order spherical aberration can be expected at such a large value of third order. It appears that the high order amounts to about -0.005, but it will grow rapidly at larger values of ρ.

An alternate way to understand how the lenses work is to use the Eqs. (2.81) to (2.85). To return to the above example, it can be seen that it is necessary to find some source of positive astigmatism in order to correct the inward-curving astigmatism and field curvature shown in Fig. 39. Equation (2.83) shows that the astigmatism always has the same sign as the spherical on any surface. The task then is to find a surface that introduces positive spherical aberration. This is done by bending the lens to the left as shown in Fig. 40. This causes an even stronger second surface which increases the negative spherical for the total lens. However, the stop position can be chosen to have the chief ray pass through the second surface at close-to-zero angle of incidence. This reduces the astigmatism contribution from the surface, and the total lens can have a net positive astigmatism which can be used to cancel negative Petzval field curvature.

In the conceptual stages of design, one tends to use the thin-lens equations to guide one's thoughts, but in the later phases of design, the surface contributions are used more than the thin-lens equations. This is because the thicknesses of the lenses have significant effects.

2.6.4. Further Design Tips

The surface-contributions Eqs. (2.81) to (2.85) show that the magnitude of the aberrations is associated with the angle of incidence on the surface. One technique of design is to try to arrive at a third-order correction with all the angles of incidence being as small as possible. Large aberrations are sure to cause large aberrations for the third- and high-order aberrations. Designs that cancel large aberrations of one sign against the opposite sign lead to more sensitive designs. Even if the designer succeeds in achieving a fine balance of aberrations, the lens will be more sensitive, and the balance may not actually take place in manufacturing. It is often advisable to use extra elements in order to avoid large angles of incidence. The extra cost of the element will be compensated for by looser manufacturing tolerances. An accurate drawing of the lens showing the extreme rays passing through the lens is a significant tool used by designers. The lens should look well balanced. Sharp curves with large bendings of the rays should raise a flag of caution.

Practical designs do, however, have to balance high-order aberrations. When a surface introduces a large amount of third-order aberration, it should be corrected with a surface as close by as possible. It is difficult to achieve a good balance if the high-order aberrations allow the rays to become too far separated from the third-order predictions. This is because there are many more high-order aberrations, which are not correctable one at a time. If they are eliminated as close as possible to their source, then there is a tendency to correct the complete assembly of aberrations. If the correcting is done at a remote position, a different distribution of correction occurs.

2.7. Optical Design by Computer

The balancing of high-order corrections is difficult for a human, but automatic design programs are effective in doing this. It is up to the designer, however, to provide the appropriate surfaces and indices of refraction to enable programs to find a practical solution.

Note that the coma Eq. (2.82) depends on $\bar{i}i$. This means that even when S is of the same sign on two surfaces, the coma (σ_2) can cancel if on one the $\bar{i}$ reverses its signs. This is why lenses work best when the aperture stop is in the center of the lens. The lenses on opposite sides of aperture can then have opposite values of coma. Coma is a bad characteristic in a lens, for it introduces an unsymmetrical image which is easily recognized as a defect. It should be reduced to the minimum possible value.

There are several commercial ray-tracing and automatic target-seeking programs for lens design. These are:

1. ACOSS. Supported by Scientific Calculations, 7796 Victor-Mendon Road, Victor, New York 14564.
2. CODE V. Supported by Optical Research Associates, 550 N. Rosemead Blvd., Pasadena, California 91107, (213) 798-0766.
3. COOL GENII. Supported by Genesee Computer Center, 20 University Avenue, Rochester, New York 14605, (716) 232-7050.
4. OSLO. Supported by Sinclair Optics, 6780 Palmyra Rd., Fairport, New York 14450.
5. LASL. A Los Alamos program, which is available to public users. Inquiries may be directed to Berlyn Brixner or Morris Klein, Los Alamos, National Laboratories, Los Alamos, New Mexico 87545.
6. Kidger Optics Ltd. Pavane House, Church Rd. Crowborough, East Sussex TN6 1BL. England.
7. Sciopts Enterprises. San Jose, CA.

This chapter has been in process for several years. As a result several changes have occurred. The present list of Design Programs, however, is lacking in one major program which must be added.

8. Synopsys—Optical Systems Design Inc., P.O. East Bothbay, Maine 04544-0247.

These are sophisticated programs, which have been under development for over 20 years. In their present forms, they can handle almost any geometrical optics problem. They are available through time-sharing or remote-batch terminals and can be reached from any place with telephone service. The OSLO, Genii, Kidger, and Sciopts programs are written for desktop calculators.

The programs are by no means push-button design programs. It takes an experienced person with considerable design background to use them properly. The programs are so flexible that there are many ways to make foolish mistakes; so computing time is cheap for the wise and expensive for the inept. These programs are not recommended for anyone who is not prepared to devote the time of several months to learn how to design with them. In addition, the learner needs to have some guidance from an experienced user of the program. In the hands of an amateur, the programs are far better at maximizing cost than in minimizing image errors and reducing manufacturing sensitivities.

There are numerous programs now being developed for small personal computers. These have their place and undoubtedly will become popular for technical people wishing to engage in part-time lens design. Designing lenses is really a great deal of fun and ranks with chess as a challenging intellectual activity. Competing in the design of various lenses is now devoid of drudgery. However, when it comes to serious design, one has to remember that a paper design is just that. If it is to be turned into a mounted and tested lens, the paper design initiates a large amount of expensive engineering and must involve careful trade-off decisions and tolerance analysis. This soon stops to be fun and turns into a demanding consideration of all the steps that have to go into building the lens and mount. To do this well and not to depend upon other people's experience and good judgement requires a great deal of analysis. Any designer will tell you that there never comes a time when he or she would not like to have some more time and computing data to help make some of the decisions needed to complete the design task. It is always that the budget is overrun, other jobs are more pressing, or the deadline is tomorrow morning. If large computers cut the cost of computing by a factor of ten, the lens designers will find ten times as many things they would like to evaluate. It is hoped that the small-computer fascination will not reduce the effort going into the large design packages, for they need to be faster, easier to use, and much more economical.

Appendix

1. *Definition of an Optical Surface.* The general expression for an optical surface with symmetry of revolution about the optical axis z is given by[9]:

$$z = \frac{cS^2}{1+[1-(k+1)c^2S^2]^{1/2}} + A_1S^4 + A_2S^6 + A_3S^8 + A_4S^{10}, \quad \text{(A2.1)}$$

where $S^2 = x^2 + y^2$ and $c = 1/r$, with r being the radius of curvature at the vertex. The quantities A_1, A_2, A_3, and A_4 are the aspheric deformation constants, and k is a function of the eccentricity of a conic surface given by $k = -e^2$ and is called the *conic constant*. If all the coefficients A_i are zero, the surface is a conic of revolution, or an oblate spheroid, according to the value of K, as follows

hyperboloid: $\qquad k < -1$
paraboloid: $\qquad k = -1$

prolate spheroid, or ellipsoid, obtained by
rotating an ellipse about its major axis: $\quad -1 < k < 0$
sphere: $\quad k = 0$
oblate spheroid, obtained by
rotating an ellipse about its minor axis: $\quad k > 0$

The aspherical or conic surface may also be represented by a sphere with radius of curvature equal to that of the vertex of the aspherical surface (osculating sphere), plus some deformation terms, as follows

$$a = \frac{cS^2}{1+\sqrt{1-c^2S^2}} + B_1 S^4 + B_2 S^6 + B_3 S^8 + B_4 S^{10}, \qquad (A2.2)$$

where

$$B_1 = A_1 + \frac{[(k+1)-1]c^3}{8}, \qquad (A2.3)$$

$$B_2 = A_2 + \frac{[(k+1)^2-1]c^5}{16}, \qquad (A2.4)$$

$$B_3 = A_3 + \frac{5[(k+1)^3-1]c^7}{128}, \qquad (A2.5)$$

$$B_4 = A_4 + \frac{7[(k+1)^4-1]c^9}{256}, \qquad (A2.6)$$

Spherocylindrical surfaces[10] with sagittal curvature c_s and tangential curvature c_t may be represented by:

$$z = \frac{c_s x^2 + c_t y^2}{1+[1-\frac{(c_s x^2 + c_t y^2)^2}{(x^2+y^2)^{1/2}}}. \qquad (A2.7)$$

This expression reduces to a spherical surface when both curvatures, sagittal and tangential, are equal.

2. *Ray Tracing in Aspherical Surfaces.* The equations for tracing skew rays through an aspherical surface with rotational symmetry have been described by Hopkins and Hanau.[11] This procedure will not be repeated here, but can be summarized by the following five steps, where the initial ray data are $x_{-1}, y_{-1}, z_{-1}, K_{-1}, L_{-1}, M_{-1}$. (See Fig. 41)

(1) The intersection of the ray x_T, y_T with the plane tangent to the vertex of the surface must first be calculated as follows

$$\frac{d_{-1}}{n_{-1}} = (t_{-1} - z_{-1})\frac{1}{M_{-1}}, \qquad (A2.8)$$

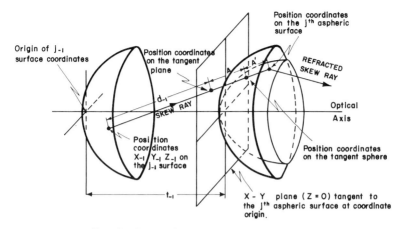

FIG. 41. Ray tracing through aspherical surfaces.

$$y_T = y_{-1} + \frac{d_{-1}}{n_{-1}} L_{-1}, \qquad (A2.9)$$

$$x_T = x_{-1} + \frac{d_{-1}}{n_{-1}} K_{-1}. \qquad (A2.10)$$

(2) Next, the intersection with the osculating sphere is calculated:

$$H = c(x_T^2 + y_T^2) \qquad (A2.11)$$

$$B = M_{-1} - c(y_T L_{-1} + x_T K_{-1}) \qquad (A2.12)$$

$$n_{-1} \cos I = n_{-1} \left(\left(\frac{B}{n_{-1}} \right)^2 - cH \right)^{1/2} \qquad (A2.13)$$

$$\frac{A}{n_{-1}} = \frac{H}{B + n_{-1} \cos I} \qquad (A2.14)$$

$$x = x_T + \frac{A}{n_{-1}} K_{-1} \qquad (A2.15)$$

$$y = y_T + \frac{A}{n_{-1}} L_{-1} \qquad (A2.16)$$

$$z = \frac{A}{n_{-1}} M_{-1} \qquad (A2.17)$$

(3) The vector normal to the surface, whose cosine directors are directly proportional to U, V, and W, is found by means of the gradient. Simul-

taneously, the intersection x_n, y_n with the aspherical surface is also calculated by an iterative method until the error $\Delta A'$ is sufficiently small. The subscript n stands for the order of approximation.

$$S_n^2 = x_n^2 + y_n^2, \tag{A2.18}$$

$$W = [1 - c^2 s^2]^{1/2}, \tag{A2.19}$$

$$F = z - \left[\frac{cS^2}{1+W} + B_1 S^4 + B_2 S^6 + B_3 S^8 + B_4 S^{10}\right], \tag{A2.20}$$

$$E = c + W[4B_1 S^2 + 6B_2 S^4 + 8B_3 S^6 + 10B_4 S^8], \tag{A2.21}$$

$$U = -XE, \tag{A2.22}$$

$$V = -YE, \tag{A2.23}$$

$$\frac{A'}{n_{-1}} = \frac{-FW}{K_{-1} U + L_{-1} V + M_{-1} W}, \tag{A2.24}$$

$$x_{n-1} = x_n + \frac{A'}{n-1} K_{-1}, \tag{A2.25}$$

$$y_{n-1} = y_n + \frac{A'}{n-1} L_{-1}, \tag{A2.26}$$

$$z_{n-1} = z_n + \frac{A'}{n-1} M_{-1}. \tag{A2.27}$$

(4) Finally, the director cosines for the refracted ray are calculated by using Snell's law of refraction.

$$G^2 = U^2 + V^2 + W^2, \tag{A2.28}$$

$$G_{n-1} \cos I = K_{-1} U + L_{-1} V + M_{-1} W, \tag{A2.29}$$

$$Gn \cos I' = n\left(\left(G\frac{n_{-1}}{n} \cos I\right)^2 - G^2 \left(\frac{n_{-1}}{n}\right)^2 + G^2\right)^{1/2}, \tag{A2.30}$$

$$P = \frac{Gn \cos I' - Gn_{-1} \cos I}{G^2}, \tag{A2.31}$$

$$K = K_{-1} + UP, \tag{A2.32}$$

$$L = L_{-1} + VP, \tag{A2.33}$$

$$M = M_{-1} + WP. \tag{A2.34}$$

If the surface is spherocylindrical, Eqs. (A2.20) to (A2.23) are replaced by the following[10]:

$$Q = c_s x^2 + c_t y^2, \qquad (A2.35)$$

$$R = [1 - Q^2/S^2]^{1/2}, \qquad (A2.36)$$

$$F = z_n - \frac{Q}{1+R}, \qquad (A2.37)$$

$$U = W \frac{-\left[2c_s x_n(1+R) + 2\left(\frac{c_s x_n^2}{S^4}\right)(2c_s - Q)Q\right]}{(1+R)^2}, \qquad (A2.38)$$

$$V = W \frac{-\left[2c_t y_n(1+R) + 2\left(\frac{c_t y_n^2}{S^4}\right)(2c_t - Q)Q\right]}{(1+R)^2}. \qquad (A2.39)$$

References

1. R. Kingslake, *Applied Optics and Optical Engineering*, Vol. 3, Chap. 7, p. 269, Academic Press, New York, 1965.
2. R. S. Longhurst, *Geometrical and Physical Optics*, 3d ed., Longmans, New York, 1973.
3. D. P. Feder, "Optical Calculations with Automatic Computing Machinery," *J. Opt. Soc. Am.* **41**, 630 (1951).
4. W. T. Welford, *Aberrations of the Symmetric Optical System*, Academic Press, New York, 1974.
5. W. T. Welford, *Geometrical Optics*, North-Holland Publ., Amsterdam, 1962.
   This is an excellent book for undergraduate teaching of optics for physics and engineering students. It could be used in the sophomore or junior year.
6. C. G. Wynne, "Secondary Spectrum with Normal Glasses." *Opt. Commun.* **21**, No. 3 (1977).
7. C. G. Wynne, "A Comprehensive First-order Theory of Chromatic Aberration. Secondary Spectrum Correction without Special Glasses," *Optica Acta* **25**, 627 (1978).
8. H. Buchdahl, *Introduction to Hamiltonian Optics*, Cambridge Univ. Press, Cambridge, 1970.
9. D. Malacara, *Optical Shop Testing*, p. 479, Wiley, New York, 1978.
10. C. Menchaca and D. Malacara, "Toroidal and Spherocylindrical Surfaces," *Appl. Opt.* (In Press.)
11. R. E. Hopkins and R. Hanau, in *Military Standardization Handbook: Optical Design*, *MIL-HDBK 141*, U.S. Defense Supply Agency, Washington, D.C., 1962.

3. THE COMPONENTS IN THE BASIC OPTICAL SYSTEMS

R. E. Hopkins

Optizon Corporation
300 East River Rd.
Rochester, New York 14623

3.1. The Components in the State-of-the-Art Optical Systems

In this chapter, the components used in the most common imaging systems will be described. The performance of the individual components will be explained well enough to help the reader select the appropriate type of lens to use in a laboratory setup for trying out as a prototype of a new instrument. Most of the designs shown were designed for a particular requirement, so any change of it may require a new optimized design. However, some requirements are not lens-limited so they may be usable as they are. It is hoped that this chapter will provide benchmarks in the confusing world of optical components which have been developed over one hundred years.

3.2. The Terrestrial Telescope

A good example of a terrestrial telescope is the optics in a pair of prism binoculars. The optics found in binoculars have been made in large numbers, and many shops have the capability of making this kind of optics. Surplus lenses can be found at reasonable prices. Figure 1 is a designers view of one of the pair of the optics in a 10×50 binocular. The focal length of the objective is 254 mm. The eyepiece is an Erfle eyepiece with a focal length of 25.4 mm. The drawing shows the path of the axial ray as it passes through the system. It shows the path of the lower and the central chief ray of the oblique beam. Notice that the full upper ray is vignetted by the lenses of the eyepiece. This is done to keep the size of the eyepiece and prisms smaller. There is a reduction of beam diameter and a loss of field brightness if the pupil of the eye is a full 5 mm, but it is not noticeable. It is quite common to allow as much as 50% vignetting. This means the area of the beam may be reduced by 50%. This telescope has a 5-mm exit pupil, which is a bit larger than the pupil of the eye in normal daylight, but the eye pupil

60 THE COMPONENTS IN THE BASIC OPTICAL SYSTEMS

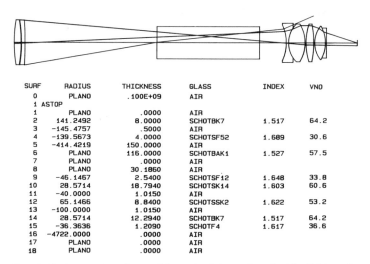

SURF	RADIUS	THICKNESS	GLASS	INDEX	VNO
0	PLANO	.100E+09	AIR		
1 ASTOP					
1	PLANO	.0000	AIR		
2	141.2492	8.0000	SCHOTBK7	1.517	64.2
3	-145.4757	.5000	AIR		
4	-139.5673	4.0000	SCHOTSF52	1.689	30.6
5	-414.4219	150.0000	AIR		
6	PLANO	116.0000	SCHOTBAK1	1.527	57.5
7	PLANO	.0000	AIR		
8	PLANO	30.1860	AIR		
9	-46.1467	2.5400	SCHOTSF12	1.648	33.8
10	28.5714	18.7940	SCHOTSK14	1.603	60.6
11	-40.0000	1.0150	AIR		
12	65.1466	8.8400	SCHOTSSK2	1.622	53.2
13	-100.0000	1.0150	AIR		
14	28.5714	12.2940	SCHOTBK7	1.517	64.2
15	-36.3636	1.2090	SCHOTF4	1.617	36.6
16	-4722.0000	.0000	AIR		
17	PLANO	.0000	AIR		
18	PLANO	.0000	AIR		

FIG. 1. Designers view of one of the pair of the optics of a 10×50 binocular.

tends to open up when the binoculars are placed into use. The eye then receives less light and the iris expands to adjust to the lower light level.

The objective can only be corrected for spherical aberration, coma, and axial color (Tach Eq. 2.54). The prism adds -0.88 waves of positive spherical aberration, and the eyepiece adds -0.21 waves of negative spherical aberration. The objective adds 0.49 waves of negative spherical aberration. This tends to be the case, and an objective that has been designed for use with a prism will be undercorrected for spherical aberration. The same is true for axial color. The prism adds positive Tach, so the objective is slightly undercorrected for color.

The two most serious aberrations in a telescope of this type are the Petzval field curvature and the astigmatism. Both, the objective and the eyepiece cannot be corrected well for Petzval curvature. The objective Petzval radius in this case is 1.45 times its focal length. The eyepiece Petzval radius is 1.485 times its focal length. The eyepiece is the big offender, because its focal length is 10 times shorter than the objective. The best that can be done, then, is to use the eyepiece to adjust the astigmatism to the best possible compromise. Unfortunately, this is where the art in lens design comes in. There are many conflicting things to think about. The eye can accommodate as the observer looks at different points in the field; there always will be some astigmatism, and sometimes there is a reticle in the focal plane which is used for range estimating.

It is far beyond the scope of this section to go into all the details of how good to leave the correction of the off-axis aberrations. The uninitiated will perhaps be surprised to see how bad the imagery off axis is, and the number of designers who would agree with the selected balance of aberrations would be about one. The following, however, should be said. Visual instruments like the above are different from most other optical systems. The telescope's capability should concentrate on the quality of the central image. This image should be made as good as possible because of the way a pair of binoculars is used. The viewer observes objects in the center of the field and does not turn his or her eyes with respect to the instrument in order to see things off the axis. It is instinctive to turn the binoculars and the head to see points at the edge of the field.

The main purpose of the wide field is to reduce the feeling of looking down a narrow tunnel and to allow one to detect any motion in the scene. Trying to improve the performance of the edge of the field image should not be done if it detracts from the central image and adds to the complexity and cost.

There is one further interesting point about the wide-angle binocular system. There is 9% distortion in the telescope. This amount of distortion is hardly noticeable, for the field of view is round. The apparent field angle of the telescope is 30°. The ratio of the apparent field to the real field is then 10×. Note that the distortion causes the apparent field to follow the theta law rather than the tangent law, as it would if there were no distortion. This is actually the way it should be because, as stated above, one uses the telescope by turning the telescope in order to see different objects in the field. Since the apparent field follows the theta law, all the objects in the apparent field move with the same angular velocity as the observer scans the real field by turning the binoculars. If the distortion is corrected to zero, the objects at the edge of the field would appear to move faster than the objects in the center of the field. This causes some confusion to the viewer. The telescope should therefore not be corrected for distortion. This is convenient, because correcting the distortion in an eyepiece is difficult.

Wide-field binocular telescopes usually have eyepieces with apparent fields between 30 and 35°, and most of them use the Erfle or similar type eyepieces.

3.2.1. More on Eyepiece Design

Eyepieces with apparent fields up to 45° have been made using an aspheric surface. The aspheric surface is used mainly to correct the spherical aberration in the position of the exit pupil. Spherical aberration of the exit pupil makes it difficult to position the eye, so that the entire field is seen at once. When there is spherical aberration of the exit pupil, shadows will appear as the eye rotates to see objects near the edge of the field.

There are many other types of eyepieces,[1,2] and a few are shown in Fig. 2 along with some of their characteristics. All the eyepieces have a focal length of 2.5 cm, with γ being the ratio of the Petzval radius to the focal length. The eye relief is the distance from the exit pupil to the first surface of the eye lens. Note that it is less than the focal length of the eyepiece. The eye relief for all these eyepieces were chosen assuming that they were to be used with a telescope objective of 254-mm focal length. The resulting telescope would have a magnifying power of $10\times$. The Huygens eyepiece is still used in low-power telescopes and microscopes. The Ramsden eyepiece is used in narrow-field telescopes and microscopes when a cross hair is needed. The Kellner eyepiece is used as an upgrade for the Ramsden eyepiece. It is also called an achromatic Ramsden. The symmetrical eyepiece is a popular design for rifle scopes. It is sometimes called the Plossl eyepiece. The orthoscopic eyepiece is commonly used for surveying telescopes. It and the symmetrical eyepiece have long-eye relief, which is needed for short-focal-length eyepieces.

Eyepieces are really not useful for applications in optics other than as eyepieces in visual instruments. The individual lenses are not good for collimating or focusing lenses, for they are large for their focal length, and the aberrations are large. Some of the surplus houses take eyepieces apart and sell the individual components. They are not useful for anything other than magnifiers.

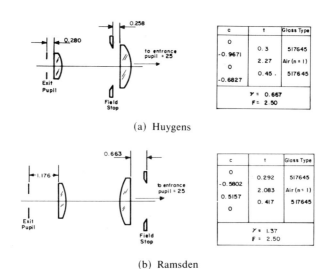

FIG. 2. Some eyepieces and their characteristics. The quantity γ is the ratio of the Petzval radius of curvature to the effective focal length.

THE TERRESTRIAL TELESCOPE

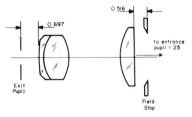

(c) Kellner

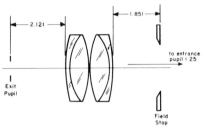

(d) Symmetrical

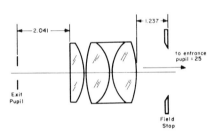

(e) Orthoscopic

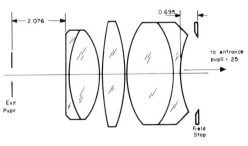

(f) Erfle

FIG. 2. Continued.

3.2.2. Other Telescope Objectives

The objectives in telescopes are often useful in the laboratory as focusing or collimating lenses. If one is lucky, nearly perfect lenses can be found at bargain prices. The objectives from binoculars and small telescopes are usually cemented. This makes them much easier to mount in an experiment. Some of the larger telescope objectives are air-spaced. Lenses of this type can be made to provide superior performance, but they must be mounted carefully or the design performance will not be realized. In the section on laser-focusing and collimating lenses, the detailed performance of these lenses will be described in more detail. Telescope objectives form diffraction-limited performance over extremely small fields. Beyond half a degree off axis, the lens will start to show astigmatism.

Some of the telescopes used in modern surveying equipment use higher-speed objectives (smaller F/number or larger NA) than the objective shown in Fig. 1. For a given magnifying power and aperture diameter, this results in a shorter and more compact telescope. This introduces secondary spectrum, so more color aberration will be evident to the user. The shorter-focal-length objective also leads to a shorter focal length for the eyepiece, and the eye relief may cause discomfort.

3.2.3. Prism Designs

The Porro Type 2 prism system is by far the most common prism used in binoculars see (Fig. 3). They provide an offset between the objective separation and the eyepeice separation. The increased separation of the objectives is sometimes said to enhance the stereo baseline, but actually this is an extremely small effect. The increased separation of the objectives does, however, provide some more stability, which is useful in high-power binoculars. The Porro Type 2 prisms consist of two 45–45–90-degree prisms,

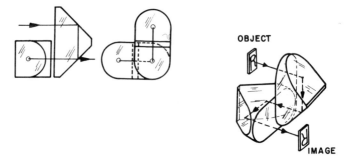

FIG. 3. The Porro Type 2 prism system used in binoculars.

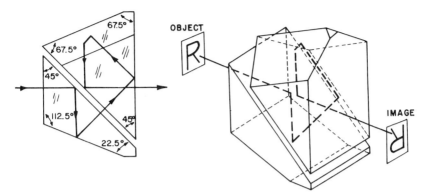

FIG. 4. The Pechan–Schmidt prism used in some small terrestrial telescopes or binoculars.

and the angular tolerances are around one minute. This means that they cannot be used as roof prisms in other applications.

The Pechan–Schmidt prism, shown in Fig. 4, can also be used in binoculars. This is a much more expensive prism system than the Porro because it uses a roof prism. A good roof prism must have the angle on the roof, equal to 90° plus or minus a second. The roof edge must also be sharp and free from nicks and scratches, for they will appear in the visual field. Roof prisms make compacted erecting systems. There are some excellent but expensive binoculars using this prism.

The Pechan–Schmidt prism should not be confused with the Pechan prism. The Pechan prism looks like the Pechan–Schmidt prism but has no roof edge. As the light passes though it, there are only five reflections, which is an odd number, so that it forms a left-handed image. When the optical system is made up of a series of rotationally symmetric lenses, it must have an even number of reflections to maintain a right-handed image. A telescope which uses prisms to erect the image must have an even number of reflections. By making the top surface of the Pechan into a roof, it becomes a Pechan–Schmidt with six reflections. The Pechan prism is used to rotate an image around the instrument's optical axis, but it must be inserted into an optical system which has an odd number of reflections. Sometimes the surplus houses misname these prisms, so it is necessary to check as to the number of reflecting surfaces. The two components of these Pechan prisms are sold as separate units. They are not easily mounted together. They resemble watermelon seeds, and so it is best to leave the mounting up to optical experts. Buying the assembled unit is recommended.

The names of prisms have become mixed up over the years they have been used. Figure 5 gives a selection of the most commonly used prisms, but there are many others for special applications.

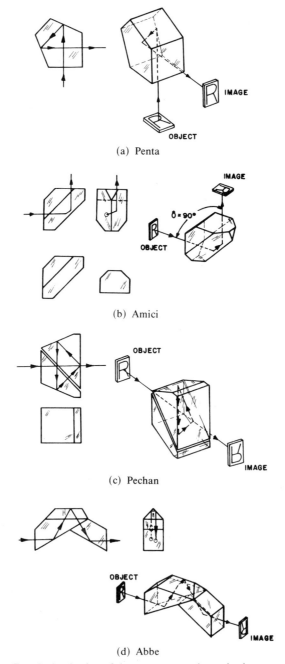

(a) Penta

(b) Amici

(c) Pechan

(d) Abbe

FIG. 5. A selection of the most commonly used prisms.

There is always the question as to whether to use prisms or mirrors. Prisms have the following advantages over mirrors:

1. Once prisms are made, they maintain their structural integrity.
2. It usually is possible to use total reflection.
3. A prism has an apparent thickness which is the geometrical thickness divided by the index of refraction of the glass. A Porro prism can be inserted in a space where there would not be enough space to use mirrors. Many people have proposed using mirrors in place of the Porro prism in order to make a light-weight pair of telescope spectacles. The mounting of the mirrors is such a difficult task that it has never been a practical approach.

Mirrors have the following advantages:

1. They are light-weight.
2. For large sizes, the cost of prisms becomes prohibitive.
3. With modern high-reflecting coatings, they are slightly higher-reflecting than a prism with total reflection. This is because the prism requires two outer surfaces with reflection loses.

3.3. Microscope Optics

The simplest microscope is the common ordinary magnifier. The principle of the magnifier is illustrated in Fig. 6. To view an object more closely and

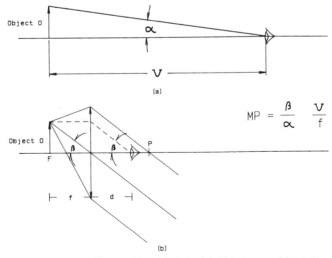

FIG. 6. A single-lens magnifier working principle. (a) Object seen without the magnifier, at the nearest viewing distance V. (b) Object seen with the magnifier.

make it appear enlarged, an observer brings the object as close as the observer can focus. This distance is called the nearest distance of distinct vision. To view from this distance, it is necessary to use the full focusing capability of the eye (accommodation of the eye). This nearest viewing distance is called V. The object being observed then subtends the angle α. V differs from person to person, but the standard observer has a V equal to 250 mm or 10 inches. A farsighted person has a V greater than 250 mm, while a nearsighted person has a V shorter than 250 mm. A single lens acts as a magnifier as illustrated in Fig. 6(b). When the object is located at the first focal point of the lens, the light from the object is collimated, so that the object appears to subtend the angle β for any position between the lens and the position P. The eye has to be located within the upper and lower rays emerging from the lens. If the lens has aberrations, the emerging rays will not all be parallel, so that the angle will not be constant and there will be distortion of the image. For this reason, it is best to place the eye as close to the axis of the lens as possible. The magnifier, when used this way, provides magnifying power (MP) given by the equation,

$$\text{MP} = \tan \beta / \tan \alpha. \qquad (3.1)$$

The equation states that the MP of a magnifying lens is the ratio of the tangent of the visual angle subtended by the object when using the magnifier, divided by the tangent of the angle subtended by the object when viewed from the distance V. As can be seen from Fig 6, the magnifying power is given by the equation

$$\text{MP} = \frac{V}{f}. \qquad (3.2)$$

Magnifiers are usually labelled according to their MP, using the convention that the value of V is 250 mm. A $10\times$ magnifier then can then be assumed to have a focal length of 25 mm. One should use a magnifier so that the virtual image is viewed with the eye in the relaxed condition. Then the virtual image appears to be at the observer's far point. The way to do this is as follows. Place the magnifier on the material to be viewed. Then move the magnifier away from the material until the virtual image becomes blurred. Finally move the magnifier closer until the image first becomes sharp. This procedure positions the magnifier so that the virtual image is located at the most restful position for the observer. If the observer's far point is at infinity, the virtual image will also appear the same size for any distance d that the eye is placed behind the magnifier. As d becomes smaller, the field of view will increase. The magnifying power of a given lens differs from person to person. An older person receives more benefit than a younger person. The older person's near point of distinct vision has receded because of loss of accommodation, so the value of V may be 500 mm. A lens with a focal length of 100 mm will provide the normal young observer a magnifying power of $2.5\times$, while a person with V equal to 500 mm will obtain $5\times$ power. Likewise, a farsighted person will gain much more magnifying power than a nearsighted person, because his/her V is larger.

Magnifiers range in power from about $2\times$ to $20\times$. The common reading glasses run from $2\times$ to $5\times$ and are single lenses. The best shape lens for a low-power reading glass is plano-convex. When the observer places the eye at a considerable distance from the lens, the lens should be used with the plano surface facing the object. However, if the eye is placed close to the lens, it should be used the other way around. Since most observers do not realize this, magnifiers are made equiconvex, and there is some pincushion distortion when the eye is placed close to the lens. Many of the reading magnifiers are made out of plastic, and one or both of the surfaces are made aspheric to help reduce the distortion.

People with impaired vision use telescope spectacles (see Fig. 7). These are basically small lens magnifiers of low power and long working distance backed up by a Galilean telescope. These are made small and light-weight and some instrument makers use them for detail work.

Magnifiers of higher power ($5\times$ to $20\times$) are common, but they are more tedious to use, for the eye has to be located close to the lens because the focal length is short and the diameter of the lens becomes smaller with increased power. Most high-power magnifiers are also made symmetrical so that the user can use them either side up. The Hasting triplet is probably the most common $10\times$ magnifier. Any of the eyepieces used in binoculars or telescopes make good magnifiers.

In order to achieve higher magnifying powers, it is necessary to go to the compound miscroscope which consists of an objective that forms an enlarged image of the object that can be viewed with an eyepiece. The

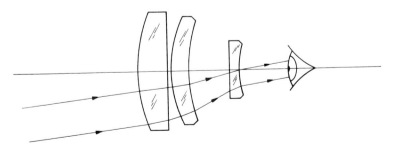

FIG. 7. Galilean telescope to help people with impaired vision.

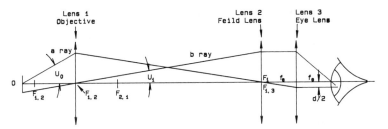

FIG. 8. The Gaussian optics of a microscope.

principle is illustrated in Fig. 8. The objective images the object with magnification to the focal plane of the eyepiece. The eyepiece has a field lens and an eye lens which collimates the light so that the enlarged virtual image of the object appears at infinity. The eyepiece images the objective to the exit pupil, which is the position for the eye doing the observing. The magnifying power of the microscope is then given by the equation:

MP = m of the objective × MP(the magnifying power of the eyepiece)

$$m = \frac{(nu)_0}{(nu)_1} = \frac{\text{NA}}{u} = \frac{-x'}{f} \quad (3.3)$$

It follows that:

$$d = -2\,\text{NA}\,\frac{V}{\text{MP}} \quad (3.4)$$

where NA = nu, and in a well-corrected objective,

NA = $n \sin U_0$,

and d is the diameter of the exit pupil.

Equation (3.4) shows that the diameter of the exit pupil varies inversely with the magnifying power of the microscope. The normal eye-pupil diameter for optimum resolving power is 2 mm. By inserting 2 mm for d in Eq. (3.4), it follows that

Optimum MP = 250 NA (3.5)

In the past it has been standard practice among microscope makers to use 160 mm as the distance from the second focal point of the objective to the first focal point of the eyepiece. The tube lengths were made so that when the objective was screwed in place, the tube length was set at the standard length. The eyepiece would be positioned in the correct position to maintain the 160-mm distance. Since x' is fixed, the magnification of the objective is directly related to its focal length. For example, if the focal length of the objective is 16 mm, it should magnify the object 10 times when it is positioned correctly in the microscope. The manufacturers began engraving the magnification directly on the objective. A 10× microscope objective would then be a 16-mm objective. An 8-mm objective would be marked as 20×. Today the focal length of micro-objectives are almost never marked with the focal length, and sometimes the 160-mm condition is not used. Some microscope objectives use 250 as the value of x'. This means that whenever a microscope objective is to be used to make precise dimensional measurements, one cannot use the value of the magnification engraved on the lens, unless it is used at the precise conjugates that the manufacturer intended. If this distance is not known, then it is necessary to measure the magnification by using a fine scale in the object plane and a filar micrometer in the eyepiece. There is another important reason for using microscope objectives at the correct tube length. The lens was designed for a given tube length, and its performance will degrade at different conjugate distances. This is most noticeable in high-power objectives.

Biological objectives above 20× are designed for use with thin-glass coverslips. The thin plates of glass introduce overcorrected spherical aberration which the designer has corrected for. If the lens is used without the coverslip but at the right conjugates, it will be undercorrected for spherical aberration. It can be used at a longer tube length to improve the spherical correction, but this must be done by testing the image quality. The longer tube length will change the magnification from the marked value. If a biological objective is to be used at its correct tube length, then a 0.18-mm-thick coverslip should be used for optimum correction of spherical aberration.

Equation (3.4) shows that the diameter of the exit pupil varies inversely with the MP of the microscope. The manufacturers have designed their standard series of objectives to maintain the diameter of the exit pupil more or less constant. Table I lists the most common series of objectives found on compound microscopes.

TABLE I. The Optical Properties of Standard Microscope Objectives

f(mm)	m	NA	MP	d(mm)
32	5	0.125	50	1.25
16	10	0.25	100	1.25
8	20	0.50	200	1.25
4	40	0.85	400	1.06
2	80	1.29	800	0.743

Notice that the diameters of the exit pupils are all less than the normal 2-mm eye pupil. Users have found that they like more magnifying power than the theoretical limit set by the eye. This means some loss of light and that the gain in resolving power is not as great as the magnifying power suggests. It is common experience that users are comfortable with the extra magnifying power. The 40× objective would require an NA of 1 to keep the pupil size at 1.25 mm; but this is not possible in practice. An NA equal to 0.85 is just about the practical limit for an air objective. The objectives above 40× have to be oil-immersion in order to keep the diameter of the exit pupil from becoming too small. The small pupil diameter results in loss of light and a decrease of visual acuity because of diffraction. The smaller pupils require keeping the lenses of the eyepiece clear of dust. Older people will have considerable trouble with floaters in their eyes obstructing the small beam entering the eye.

The construction of similar objectives listed in Table I are shown in Fig. 9.[3] As the NA is increased, the complexity and the cost of the lenses increases. The powers above 8× require extremely careful optical design and manufacturing. The making of microscopes requires special optical processing and testing equipment. Since these are visual lenses, they have to be corrected for a broad band of wavelengths. It is necessary to correct the spherical aberration to within the Rayleigh limit (one quarter of a wave of OPD) for several wavelengths. This requires careful choice of the glasses with the optimum index and dispersions. The thickness of the lenses and

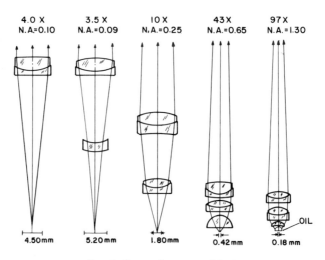

FIG. 9. Some microscope objectives.

the air spaces are all critical. Finally, the lenses have to be mounted so that the surfaces are centered on the optical axis.

The coma in microscope objectives is well corrected, but the astigmatism is not corrected completely, because the Petzval curvature of the objective is difficult to correct. There is also a relationship between the Petzval correction and the secondary spectrum. When one tries to improve the Petzval field curvature, the secondary color will increase. It is then necessary to use an unusual glass to correct the secondary spectrum. This is one of the reasons why microscope objectives use fluorite for the high-power lenses. This is expensive material and difficult to process with the precision required for diffraction-limited performance.

The field of view of microscopes is small and set by the diameter of the field stop in the eyepiece. The field stop is usually around 20 mm in diameter. The object field is the field-stop diameter divided by the magnification of the objective. The object field of a 40× objective is only about 0.5 mm. The field is by no means diffraction-limited over its entire extent. A description for the performance of lenses is the number of pixel elements it can image. This is calculated by taking the ratio of the field diameter to the diameter of the Airy disk of the lens and squaring the ratio. For example, the object field of the 40× objective is approximately 0.5 mm. The diameter of the Airy disc is given by the equation:

$$d = 1.22 \frac{\lambda}{NA} \quad (3.6)$$

The Airy disc diameter varies for the various wavelengths but averages about 0.6 micrometers divided by the NA for the visual region. For a 0.85 NA the maximum number of Airy discs in the field could then be 510,000. The whole field is not corrected this well in most microscope objectives, so the actual number of independent image points is probably less the 10,000. In the section on laser optics, some lenses will be described which can image over a million image points. The microscopes does not do as well, because it has to image a band of wavelengths and has to use the eyepiece to get the light into the observer's eye. Since the age of the modern computer, improved optical processing equipment, and new glasses, some of the modern microscopes are fairly well corrected over the entire field. Some of the new microscope designs would have appeared inconceivable 30 years ago. The prices of these objectives would also have never been anticipated. Figure 10 is a drawing of a high-power flat-field microscope objective which uses fluorite in order to reduce the secondary spectrum. (The author does not know whether this lens has been made into a commercial product, but if it has, it took some real courage.)

74 THE COMPONENTS IN THE BASIC OPTICAL SYSTEMS

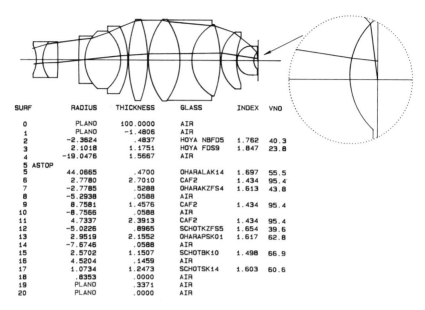

SURF		RADIUS	THICKNESS	GLASS	INDEX	VNO
0		PLANO	100.0000	AIR		
1		PLANO	-1.4806	AIR		
2		-2.3624	.4837	HOYA NBFD5	1.762	40.3
3		2.1018	1.1751	HOYA FDS9	1.847	23.8
4		-19.0476	1.5667	AIR		
5	ASTOP					
5		44.0665	.4700	OHARALAK14	1.697	55.5
6		2.7780	2.7010	CAF2	1.434	95.4
7		-2.7785	.5288	OHARAKZFS4	1.613	43.8
8		-5.2938	.0588	AIR		
9		8.7581	1.4576	CAF2	1.434	95.4
10		-8.7566	.0588	AIR		
11		4.7337	2.3913	CAF2	1.434	95.4
12		-5.0226	.8965	SCHOTKZFS5	1.654	39.6
13		2.9519	2.1552	OHARAPSK01	1.617	62.8
14		-7.6746	.0588	AIR		
15		2.5702	1.1507	SCHOTBK10	1.498	66.9
16		4.5204	.1459	AIR		
17		1.0734	1.2473	SCHOTSK14	1.603	60.6
18		.8353	.0000	AIR		
19		PLANO	.3371	AIR		
20		PLANO	.0000	AIR		

FIG. 10. A high-power flat-field microscope objective. 100 ×, NA = 0.95, full format = 0.2 mm. (Designer; Asoma Y. USP 4,505,553.)

The eyepieces for microscopes are basically simple, for the apparent fields are small and the beam is small in diameter. The achromatic Ramsden or the Huygenian are quite commonly used. Some manufacturers have eyepieces that are supposed to be used with the objective, in order to compensate for some of the achromatism in the objective. This is somewhat of a gimmick because it violates one of the principles of good design. In correcting color, it is desirable to correct the color as quickly as the aberration comes in. The colored rays should be kept together as much as possible. If the colored rays get widely separated before they are brought together, it is difficult to correct both the axial and lateral color. The objective and the eyepiece should each be achromatized.

3.4. Camera Lenses

The most widely used lenses are photographic lenses. They have been in use ever since 1840, soon after Daguerre invented his process of producing photographic images. The use of lenses in photographic cameras has been by far the largest commercial use of optics. Some camera lenses have been made in the millions-per-year range. The largest production has been for the personal camera field, but there have been many commercial uses for photographic lenses. Commercial lenses are not usually made in such large quantities as those for the personal camera, but their performance is often considerably more demanding. There is a bewildering number of lenses available today. Anyone needing a lens for some application other than for use in a camera may be able to find a bargain if an on-the-shelf lens can be found. However, the user usually knows only that a given focal length. F/number, and field of view is needed. A definition of performance is much more difficult to state. As a result, the catalogues from optical companies provide little useful information on the performance of their lenses beyond claiming that they take remarkably sharp pictures and are superior to their competitors. Hollywood cameramen were often used to endorse the popular models. Prior to 1950, the normal acceptance test was to project a set of resolution targets with the lens. Around 1960, objective measurements of lens performance began to be used for testing camera lenses. Today there is equipment available for measuring the performance of a lens, but it is an expensive operation. The problem is that lenses used in photography are theoretically capable of imaging thousands and sometimes millions of independent image points. Checking more than 5 to 10 image points in the field is expensive. A lens, even in the modern world of modulation transfer (MTF) and encircled energy, should be tested by reproducing the anticipated use as closely as possible. If the lens is to be used for taking pictures on vacation trips, the best tests is some scenes taken by the expected user. A lens to be used to print large scale integrated circuits should be tested by reproducing a mask on a wafer-integrated chip. The MTF, or the encircled energy criterion, are useful for specifying lens performance in the design phase but seldom can a lens be tested to see if it meets the performance expected, without actually using it for the intended application. The manufacturers are not anxious to describe in detail their performance, because the data can so easily be misinterpreted.

For the above reasons, the performance of lenses which are available for photography are not well understood. In the following section, a few of the most common lenses will be described.

3.4.1. The Landscape Lens

One of the oldest photographic lenses is the single-meniscus lens shown in Fig. 11. This lens is used with the aperture stop in front. It may be used at apertures around $F/10$. Lenses of the type have been used in old cameras with focal lengths up to 100 mm. The lens described in Fig. 11 was calculated for a focal length of 50 mm. It was designed to cover a standard 35-mm double-frame format (24×36 mm). The aberration curves show that the lens is undercorrected for spherical aberration, which is present in all the field

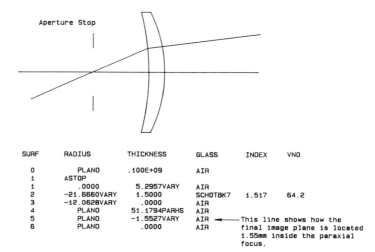

FIG. 11. The simple landscape lens. $f = 50$ mm, $F/10$, half format 21.6 mm. (Designer: R. E. Hopkins.)

positions. The Petzval radius for this lens is 1.5 times the focal length. The field curves are inward curving, so the focal plane is chosen to be 1.55 mm inside the paraxial focus which is toward the lens. The most serious aberration in this lens is the lateral color. It amounts to approximately 70 μm of difference in image height for the red and the blue rays at the edge of the field. The next most serious aberration is the spherical aberration. By focusing in from the paraxial focus, the OPD values are reduced to 1 wave for the axial focus and 3 waves for the off-axis image. This means that the imagery is determined primarily by the geometrical ray displacements. The MTF values, calculated by using diffraction analysis, for three points in the field are shown in Table II.

Using black and white photography with such a lens yields satisfactory pictures. With color film, the lateral color is evident. This lens type, simple

TABLE II. MTF Values for a Single Landscape Lens Percent MTF

Lines/mm	Axis	0.7 Field	Full field
8	69	48	36
12	49	42	18
16	30	37	8
20	15	33	4

as it is, made Kodak famous. It was used in even longer focal lengths, but several things were done to optimize the performance. The lens was vignetted heavily, and the film guide was curved into a slight cylinder in order to reduce the effect of the field curvature.

Improvement in the lens is made by using two meniscus lenses symmetrically positioned around the stop. This corrects the distortion and lateral color. The spherical aberration and astigmatism are reduced slightly, and the coma and distortion are zero by symmetry. A lens like this was made directly after World War II for a 35-mm camera. This lens and camera took beautiful pictures even in color, if there was a scene with full sunlight. The lenses were coated, so there was little scattered light. Photographic fans who had an $F/2.8$ lens and an expensive camera were considerably upset to discover that the simple double-meniscus lens had taken a picture which was superior to theirs. They had to console themselves with the fact that they could get better pictures when the light levels were not as high.

These meniscus lenses can be corrected for axial color by adding a negative lens with more dispersion than the positive lens. This leads to lenses similar to the one in Fig. 12. The lens has a 50-mm focal length and works at an $F/8$ aperture. The MTF values are given in Table III.

This lens is well corrected for the 50-mm focal length working at $F/8$, covering the double-frame 35-mm format. It could be designed to cover a

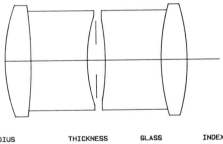

SURF	RADIUS	THICKNESS	GLASS	INDEX	VNO
0	PLANO	.9400E+08	AIR		
1	PLANO	-10.9931	AIR		
2	20.7843	3.5000	SCHOTLAK9	1.691	54.7
3	-53.2666	7.5000	SCHOTLF5	1.581	40.8
4	18.5310	1.2100	AIR		
5	ASTOP				
5	PLANO	1.2100	AIR		
6	-29.4412	7.5000	SCHOTLF5	1.581	40.8
7	25.7125	3.5000	SCHOTLAK9	1.691	54.7
8	-25.8055PARAS	.0000	AIR		
9	PLANO	37.6736PARHS	AIR		
10	PLANO	-.4818	AIR		
11	PLANO	.0000	AIR		

FIG. 12. A double-meniscus achromatic lens. $f = 50$ mm, $F/8$, half format 21.6 mm. (Designer R. E. Hopkins.)

TABLE III. MTF Values for a Double-Meniscus Achromatic Four-Element Lens

		Percent MTF			
		0.7 Field		Full field	
Lines/mm	Axis	T	S	T	S
20	65	80	77	77	73
30	43	70	67	66	61
40	25	60	59	56	52
50	13	50	51	46	44

wider angle, larger aperture, or both. The Petzval radius of this lens is −3.6 times the focal length.

Achromatic meniscus-shaped lenses are used as components in more complex lenses for modern applications. The cemented surfaces in the achromatic doublets are also used to reduce some of the higher-order aberrations by the choice of glasses that are used. The lens shown in Fig. 12 uses a high-index, low-dispersion glass (Lak9) as the positive lenses and LF5 for the negative lenses. The V-number difference between these two is only 14. With such a small difference, the cemented interfaces in the two-meniscus components have a short radius and are collective surfaces. This means that they will introduce negative spherical aberration, and the amount will increase for the off-axis ray beams. A more dispersive glass could have been used for the negative lenses, but the choice was made to keep the index low. This would increase the positive contribution to the Petzval curvature. If the negative lenses were chosen with a higher index than the positive lenses, the spherical aberration could be reduced at the sacrifice of the Petzval field flattening. This shows the case where choice of glass is important. Many of the modern designers will let the computer decide what glass to use in cemented lenses like this. The author believes that it is important to use as few glass types as possible in order to cut down on the number of glasses required of the glass manufacturers. Thus, one would select a tried and true and not too expensive high-index, low-dispersion glass for the positive lenses and then investigate the effect of several of the glasses in the negative lenses. This takes more design time, but hopefully leads to a more economical design. A characteristic of modern lenses seems to be that they require the use of several different glasses in the designs. This causes serious problems for the optical shop.

It is unlikely that four-element lenses of this type are available except in old cameras. Some of the process lenses used in photolithography use the same principle, but they usually contain more elements. They work at F/numbers from $F/3.5$ and higher.

3.4.2. The Triplet

A lens with three elements has enough degrees of freedom to correct all the third-order aberrations, provided two of them are positive or negative lenses and the third one has the opposite power. The positive lenses should be made of a low-dispersion glass, and the negative lenses should be made of high-dispersion glass. Positive lenses always form images of an object plane on a surface curved towards the lens. Negative lenses form images curved in the opposite direction. This property of lenses is expressed in the Petzval theorem (see Eq. (2.84), Section 2.6.3). The Petzval theorem states that in order to form images on a flat image plane, it is necessary to use positive and negative lenses which are separated by air spaces or by thick lenses. The design rule is stated as follows. The axial marginal ray should pass through the positive lenses at high aperture and pass through the negative lenses at low aperture. Figure 13 shows the six combinations of three lenses which could be used to correct the Petzval field curvature. The number 2 lens is the only configuration which allows correction of all the third-order aberrations and the axial and lateral color. The name *triplet* is used for the number 2 design. There probably have been more triplets designed for camera use than any other type, but today's camera and TV lenses are using more complex lenses. Several 35-mm cameras have used 50-mm focal length, $F/3.5$ triplet lenses. They cover approximately 23.5° half angle. This is a bit too much for them, but camera lenses are designed for optimum performance when they are used at about two stops down

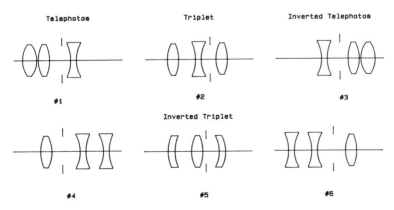

FIG. 13. Six possible combinations of three lenses to correct the Petzval field curvature.

TABLE IV. MTF Values for a Triplet Lens

Lines/mm	Axis	0.7 Field		Full field	
		T	S	T	S
20	73	41	9	51	52
30	53	25	15	39	29
40	35	15	4	31	18
50	21	11	6	25	15

from the rated F/number. Table IV lists the MTF performance for a 50-mm focal length triplet camera lens when it is operating at $F/3.5$. The lens prescription is provided in Fig. 14.

The triplets used in larger-format cameras are essentially scaled-up versions of the 50-mm focal length, and the MTF curves are not as good. A guideline used in designing camera lenses is that for optimum viewing, the observer should view the picture from the center of perspective. The center of perspective is the second nodal plane of the taking lens. A contact print from a 35-mm double-frame negative should be viewed from 50 mm in order to see the true perspective. But this is too close for most viewers who prefer to view pictures from 10 to 12 inches. Most 35-mm negatives are

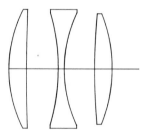

SURF	RADIUS	THICKNESS	GLASS	INDEX	VNO
0	PLANO	.1000E+09	AIR		
1	PLANO	-11.4611	AIR		
2	21.8046	3.0486	SCHOTLAK8	1.713	53.8
3	254.6790	5.4981	AIR		
4	-33.1504	1.0162	SCHOTSF8	1.689	31.2
5	21.9855	.6215	AIR		
6	ASTOP				
6	PLANO	4.6129	AIR		
7	95.9442	3.0486	SCHOTLAK8	1.713	53.8
8	-24.5733	41.4309PARHS	AIR		
9	PLANO	-.2870	AIR		
10	PLANO	.0000	AIR		

FIG. 14. The conventional triplet. $f = 50$ mm, $F/3.5$, half format 21.6 mm. (Designer: R. E. Hopkins.)

enlarged to 88 × 128 mm, which is an enlargement of 3.6 times. The center of perspective is moved out to 177 mm. Usually people view prints from about 250 mm. A normal observer can resolve approximately 6 lines/mm at this distance. This, when scaled back to the 35-mm negative size, is 24 lines/mm. The lens above does not quite do this at the edge of the field, but when stopped down a stop or two, it is adequate. Triplets used on large-format cameras will have MTF values as low as 10% at 5 to 10 lines/mm.

The triplet number 5 shown in Fig. 13 cannot be designed to be color-corrected without making the outer lenses into at least doublets. Lenses of this type will be discussed in the following section on inverted double Gauss lenses. Some of the widest-angle lenses use several negative lenses on the outside and positive lenses on the inside. Copy lenses often use extensions of this type of triplet.

3.4.3. Extensions of the Triplet

Over the years, there have been many attempts to extend the triplet to wider apertures and fields. This has been done by splitting up the power of the elements. The Tessar designer used the idea of adding a negative lens with lower-index glass and more dispersion to the positive element (see Fig. 15). The interface between the two glasses becomes a collective surface, for it bends the rays downward towards the focus. It has the effect

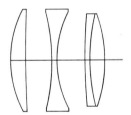

SURF	RADIUS	THICKNESS	GLASS	INDEX	VNO
0	PLANO	.1000E+09	AIR		
1	PLANO	-11.8096	AIR		
2	21.0469	3.0486	SCHOTLAK8	1.713	53.8
3	206.1837	5.3125	AIR		
4	-38.5952	1.0162	SCHOTSF8	1.689	31.2
5	20.3613	.9699	AIR		
6	ASTOP				
6	PLANO	4.6624	AIR		
7	102.4025	1.0000	SCHOTF8	1.596	39.2
8	37.3026	3.0000	SCHOTLAK8	1.713	53.8
9	-26.9139	41.2250PARHS			
10	PLANO	-.2689			
11	PLANO	.0000			

FIG. 15. The Tessar. $f = 50$ mm, $F/3.5$, half format 21.6 mm. (Designer: R. E. Hopkins.)

TABLE V. MTF Values for a Tessar Lens, $F = 50$ mm, $F/3.5$, Covering a 23° Half Angle

Lines/mm	Percent MTF				
	Axis	0.7 Field		Full field	
		T	S	T	S
20	53	36	25	67	61
30	26	21	15	46	45
40	8	13	4	26	32
50	1	9	10	11	25

of increasing the effective index of the rear element, which helps to correct the Petzval sum. Often it is used to prevent excessive overcorrection of the upper meridional rays, resulting from trying too hard to correct the Petzval sum with the middle negative lens and also resulting in high-order oblique spherical aberration. This concept is the same as described in the choice of glass for the double-meniscus lenses. The MTF values for a Tessar are listed in Table V.

The Tessar has had an illustrious reputation, but its improvement over the triplet is really not much. In the early days of lens design when there were not even any mechanical calculators, small improvements were cherished; and advertising claims, as they do today, made the most of small gains. The Tessar did show enough improvement so photographic fans could claim to see the difference. Logically, if the cemented rear element did so much in the rear, then a compound front element would help even more. The Heliar was a lens of this type, but it never received the acclaim of the Tessar. There is a great deal of art, even today, in the design of a lens for picture taking. There is always a trade-off between many lens attributes. Some people like needle-sharp images in the center of the field where the action supposedly will take place. Others like high-contrast picture over the entire field, for they do not intend to take a magnifier to look at the images. Designers have to be careful to avoid easily identifiable image errors, such as coma, lateral color, and distortion. Cost is always an important consideration.

The negative lens in a triplet can be split into two negative lenses. An interesting form of this lens is shown in Fig. 16. One of the reasons for splitting the negative lens is to provide a more symmetrical location for the diaphragm. In the triplet, it has to be located in one or the other of the air spaces, which tends to break up the symmetry. The lens shown was designed to compete with the triplet and Tessar. Its MTF performance is listed in Table VI.

CAMERA LENSES

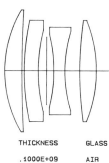

SURF	RADIUS	THICKNESS	GLASS	INDEX	VNO
0	PLANO	.1000E+09	AIR		
1	PLANO	-6.3795	AIR		
2	20.2660	3.0223	SCHOTLAK8	1.713	53.8
3	189.6286	.4012	AIR		
4	40.3416	2.4550	SCHOTSF2	1.648	33.8
5	27.1657	1.5061	AIR		
6	ASTOP				
6		1.2501	AIR		
7	-25.2943	1.9014	SCHOTSF2	1.648	33.8
8	22.6138	.0000	AIR		
9	PLANO	2.6042	AIR		
10	60.4255	2.9405	SCHOTLAK8	1.713	53.8
11	-22.3702PARHS	.0000	AIR		
12	PLANO	41.7102PARHS	AIR		
13	PLANO	-.3431	AIR		
14		.0000	AIR		

Fig. 16. The quad lens. $f = 50$ mm, $F/3.5$, half format 21.6 mm. (Designer: R. E. Hopkins.)

When the author designed this lens, he thought that the two internal negative lenses would be meniscus-shaped like the one on the left. He would then make the point that the lens was moving towards a double Gauss, and if the thickness were allowed to increase, the lens would be improved. This did not happen. It ended up with one meniscus lens and a double-concave negative lens like the one in a triplet. That is the way the ball bounces in lens design. One should always attempt to predict a solution but be prepared to be wrong. The unsymmetrical appearance of the lens is probably due to the lens being designed for infinite conjugates.

TABLE VI. MTF Values for a Quad Lens, $F = 50$ mm, $F/3.5$, Covering a 23° Half Field

		Percent MTF			
		0.7 Field		Full field	
Lines/mm	Axis	T	S	T	S
20	70	64	26	42	56
30	47	51	18	35	41
40	28	42	19	30	38
50	15	36	14	26	33

84 THE COMPONENTS IN THE BASIC OPTICAL SYSTEMS

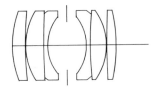

SURF	RADIUS	THICKNESS	GLASS	INDEX	VNO
0	PLANO	.5000E+08	AIR		
1	PLANO	21.6254	AIR		
2	30.2530	4.1515	SCHOTSF11	1.785	25.8
3	101.0720	.1450	AIR		
4	19.8870	5.9850	OHARASK5	1.589	61.1
5	192.2245	1.2695	SCHOTSF14	1.762	26.5
6	13.8285	6.2955	AIR		
7	ASTOP				
7	PLANO	6.2955	AIR		
8	-14.2065	1.2520	SCHOTSF14	1.762	26.5
9	103.1360	5.2085	OHARALSF15	1.804	46.6
10	-18.4460	.1450	AIR		
11	215.4830	2.9930	SCHOTSF6	1.805	25.4
12	-47.8895	36.5321PARHS	AIR		
13	PLANO	.0000	AIR		
14	PLANO	.0000	AIR		

FIG. 17. The double Gauss lens. $f = 50$ mm, $F/1.8$, half format 21.6 mm. (Designer: Takase H. USP 4,421,952.)

3.4.4. The Double Gauss Lens

The double Gauss lens can be considered as the next modification of the quad lens. The two negative meniscus lenses can be made into doublets with the negative lenses being next to stop. The negative lens should have a lower index of refraction and higher dispersion than the positive lens. This then provides an additional way of improving the Petzval field curvature. A double Gauss lens is shown in Fig. 17. This lens is the most common type of photographic lens. It is able to work at apertures as high as $F/1.5$, with field angles out to 23.5°. It is capable of covering wider field angles at reduced relative aperture. The MTF data is listed in Table VII.

A distinguishing characteristic of the lens is the way the sagittal field starts out inward curving, and then near the edge of the field, it starts to rapidly move to backward curving. The aberration curves show that there is considerable flare in the oblique aberrations which can be reduced by vignetting or by stopping down the lens. A double Gauss stopped down to $F/5$ can give nearly diffraction-limited performance. Table VIII gives the MTF values for the above double Gauss when it is stopped down to $F/5$.

Wide field-angle lenses with moderate aperatures are often called reverse double Gauss types. These lenses use the shell lenses on the outside of the lens and the positive lenses inside. Figure 18 is such a lens. The shells on the outside are negative lenses and help bring the entering and exit chief

TABLE VII. A Double Gauss US. Pat. 4,291,952 Hiroshi Takase

$F = 50$ mm, $F/1.8$, Half Field of View of 23.5°

		Percent MTF			
		0.7 Field		Full field	
Lines/mm	Axis	T	S	T	S
20	70	33	32	26	5
30	54	20	30	16	2
40	36	19	18	9	8
50	27	15	15	4	6

TABLE VIII. The MTF for the Double Gauss in Fig. 17, When Stopped Down to $F/5$

		Percent MTF			
		0.7 Field		Full field	
Lines/mm	Axis	T	S	T	S
20	87	76	71	58	76
30	78	58	48	32	61
40	70	40	27	19	51
50	62	27	14	15	43

rays around more gradually. This reduces the spherical aberration of the chief ray, so that these lenses do better at wide angles than the double Gauss lens, but they are not able to work at such large apertures. Reverse double Gauss lenses are often used in copy machines.

Extremely-wide-angle lenses are designed by extending the reverse double Gauss-type by adding extra negative lenses on the front and back of the lens. These lenses help bend the chief ray around gradually toward the central positive element, and they reverse the process after passing the aperture stop. By keeping the lens fairly symmetrical, the distortion can be corrected. In order to reduce the cosine fall-off in illumination, the designer has to design so that the aperture stop appears to expand as the field increases. When the aperture stop is viewed from the object space, the entrance pupil appears to grow in size as the lens is turned towards increasingly large field angles. If the lens is to be distortion-free, the same effect must take place as the lens is viewed from the image side. More will be said about this in the later section on illumination systems.

86   THE COMPONENTS IN THE BASIC OPTICAL SYSTEMS

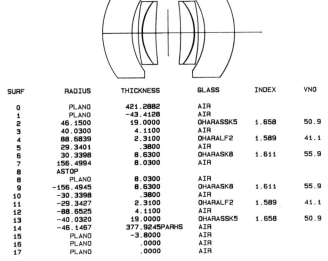

SURF	RADIUS	THICKNESS	GLASS	INDEX	VNO
0	PLANO	421.2882	AIR		
1	PLANO	-43.4128	AIR		
2	46.1500	19.0000	OHARASSK5	1.658	50.9
3	40.0300	4.1100	AIR		
4	88.6839	2.3100	OHARALF2	1.589	41.1
5	29.3401	.3800	AIR		
6	30.3398	8.6300	OHARASK8	1.611	55.9
7	156.4994	8.0300	AIR		
8	ASTOP				
8	PLANO	8.0300	AIR		
9	-156.4945	8.6300	OHARASK8	1.611	55.9
10	-30.3398	.3800	AIR		
11	-29.3427	2.3100	OHARALF2	1.589	41.1
12	-88.6525	4.1100	AIR		
13	-40.0320	19.0000	OHARASSK5	1.658	50.9
14	-46.1467	377.9245PARHS	AIR		
15	PLANO	-3.8000	AIR		
16	PLANO	.0000	AIR		
17	PLANO	.0000	AIR		

FIG. 18. The reverse double Gauss lens. $f = 211$ mm, $F/3$, half format 185 mm, magnification $= -1$. (Designer: Shinohara USP 4,490,019.)

3.4.5. Telephoto Lenses

Telephoto lenses are popular for photographic lenses, because the distance from the front surface of the lens to the image plane is less than the focal length. This makes the camera more compact. The telephoto ratio (TR) is defined as the ratio of the overall length of the lens divided by the focal length. In order to have a well-corrected lens, the Petzval surface must have a radius at least 3 times the focal length. A telephoto ratio less than the focal length requires negative power towards the rear of the lens. A thin-lens telephoto is illustrated in Fig. 19. For a given telephoto ratio, there are an infinite number of solutions for the distribution of power between the positive and the negative lens. The ratio of the power of positive and negative lenses also determines the Petzval curvature of the lens. Figure 20 is a plot of the telephoto ratio plotted against the ratio of the height of the axial ray on the second lens to the height of the axial ray of the first lens. The two lenses for this example have the same index of refraction. The curves show that there is a minimum telephoto ratio that can be attained with a given Petzval curvature. The plot also shows the effect of changing the Petzval curvature requirement. In the attempt to achieve a smaller TR, the Petzval curvature becomes more and more positive. Since a small TR causes a backward-curving field, it is practical to violate the Petzval condi-

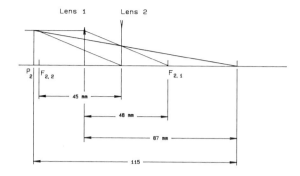

$$\text{PTZC} = -n\left[\frac{1}{f_1} + \frac{1}{f_2}\right]$$

$$= -n\left[\frac{1}{48} - \frac{1}{45}\right] = 1.5 \times .0014 = .0021$$

$$\frac{\text{PTZR}}{f'} = \frac{476.19}{115} = 4.1739$$

$$\text{TR} = \frac{\text{OAL}}{f'} = \frac{87}{115} = .76$$

FIG. 19. A thin-lens telephoto with a telephoto ratio of 0.76 with Petzval radius = 4.7f.

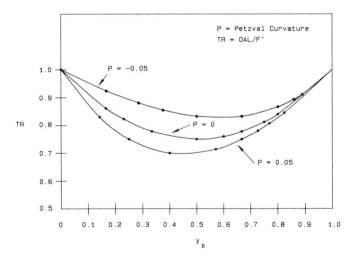

FIG. 20. A plot showing how the telephoto ratio (TR) varies with the Petzval curvature. y = axial ray height on the b lens. TR = OAL/f.

tion mentioned above in the section of the triplet. Telephoto lenses commonly use low-index glasses for positive lenses and high-index glasses for the negative lenses. The lens shown in Fig. 21 is an $F/4.5$ telephoto lens covering a half field angle of 3.1°. The telephoto TR is 0.81. The focal length is 400 mm. Note how the Petzval rule is completely violated in this lens.

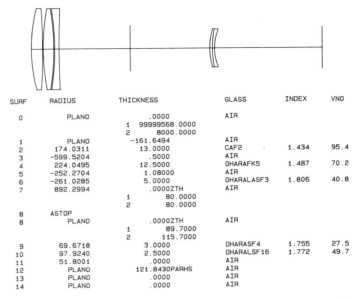

SURF	RADIUS	THICKNESS	GLASS	INDEX	VNO
0	PLANO	.0000	AIR		
		1 99999568.0000			
		2 8000.0000			
1	PLANO	−161.6494	AIR		
2	174.0311	13.0000	CAF2	1.434	95.4
3	−599.5204	.5000	AIR		
4	224.0495	12.5000	OHARAFK5	1.487	70.2
5	−252.2704	1.08000	AIR		
6	−261.0285	5.0000	OHARALASF3	1.806	40.8
7	892.2994	.0000ZTH	AIR		
		1 80.0000			
		2 80.0000			
8	ASTOP				
8	PLANO	.0000ZTH	AIR		
		1 89.7000			
		2 115.7000			
9	69.6718	3.0000	OHARASF4	1.755	27.5
10	97.9240	2.5000	OHARALSF16	1.772	49.7
11	51.8001	.0000	AIR		
12	PLANO	121.8430PARHS	AIR		
13	PLANO	.0000	AIR		
14	PLANO	.0000	AIR		

FIG. 21. A focusing telephoto. $f = 400$ mm, $F/4.5$, half format 21.6 mm. (Designer: Momiyama, USP 4,037,935.)

The positive lenses all have low index and the negative lenses have high index. Also note that the first element of the lens is made of CAF_2. This material was used to reduce the secondary spectrum which would be excessive in such a long focal length. There is a relationship between the Petzval curvature in a lens and the secondary spectrum. Whenever the Petzval curvature is made more positive, the secondary color increases. The simultaneous correction of these two aberrations are almost always in conflict. Since telephoto lenses tend to have a positive field curvature, they are also liable to have excessive secondary color. The use of CAF_2 was a bold and expensive decision for such an element. The author does not known if this lens has ever been made into a commercial lens, but it is an interesting design. One of the Fluoro Crown glasses would probably be more practical.

Telephoto lenses have one other problem. Since negative power predominates in the rear lens, it will have positive distortion. Distortion grows with the third power of the image height, and so telephoto lenses are limited in the field angle they can cover. Telephoto lenses for 35-mm cameras are popular because long focal lengths are needed for distant scenery, and the telephoto ratio keeps the overall size down. Most of the modern lenses for cameras, even if they are not strictly the positive–negative combination of

CAMERA LENSES   89

the true telephoto, are designed to have a reduced overall length. This is done by keeping the negative power more to the rear of the lens.

3.4.6. The Inverted Telephoto

A negative lens followed by a positive lens forms an inverted telephoto lens. This basic form provides a long back focus. The Petzval curvature becomes positive if the inverted telephoto ratio (ITR) is excessive. Inverted telephoto lenses tend to have negative distortion and secondary color. They often are used to obtain wide fields of view, with large aperture, when negative distortion is acceptable. Lenses of this type can cover extremely wide field angles. Fields of view of 65° half angle are common, but large negative distortion is to be expected. These lenses usually have even illumination across the format because of the distortion. Inverted telephoto lenses are excellent for wide-field applications if the distortion can be tolerated. It takes a complex lens to be distortion-free and cover a 65° field angle without serious illumination fall-off near the edge of the field. A wide-angle lens that does have negative distortion, in effect, has a shorter focal length for the off-axis imaging. This means that one cannot expect to have as good an angular resolution at the edge of the field as in the center.

Figure 22 is an inverted telephoto lens. The half field angle is 32°, and the residual distortion is −2%. This type of lens is afflicted with negative

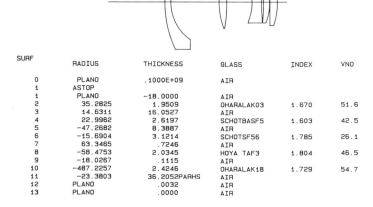

SURF	RADIUS	THICKNESS	GLASS	INDEX	VNO
0	PLANO	.1000E+09	AIR		
1	ASTOP				
1	PLANO	-18.0000	AIR		
2	35.2825	1.9509	OHARALAK03	1.670	51.6
3	14.6311	16.0527	AIR		
4	22.9962	2.6197	SCHOTBASF5	1.603	42.5
5	-47.2682	8.3887	AIR		
6	-15.6904	3.1214	SCHOTSF56	1.785	26.1
7	63.3465	.7246	AIR		
8	-58.4753	2.0345	HOYA TAF3	1.804	46.5
9	-18.0267	.1115	AIR		
10	-487.2257	2.4246	OHARALAK18	1.729	54.7
11	-23.3803	36.2052PARHS	AIR		
12	PLANO	.0032	AIR		
13	PLANO	.0000	AIR		

FIG. 22. An inverted telephoto lens. $f = 35$ mm, $F/2.8$, half format 21.6 mm. (Designer: Momiyama USP 4,037,935.)

distortion because of the preponderance of negative power in the front elements. In the design of this type of lens, it is important to allow negative distortion. If the designer insists on zero distortion, the lens field of view will have to be kept small, and the lens type has no particular advantage.

3.4.7. The Petzval Lens

The first type of triplet shown in Fig. 13 consists of two positive lenses with a negative lens in the rear. If the negative lens is placed in contact with the focal plane, it causes no telephoto effect. By spacing the two positive lenses, the lens could be called a Petzval lens with a field flattener. The two positive lenses must be color-corrected independently in order to correct the axial and lateral color. A slight adjustment is needed in the color of the two positive lenses to compensate for the color added by the negative field flattener. Lenses of this type make excellent high-speed lenses with small fields. Strictly speaking, a Petzval lens does not have the negative field lens. With just the two positive lenses, the Petzval field curvature is excessive. The second lens, however, if spaced from the first lens, can introduce positive astigmatism, and the field for the tangential rays can be flattened. Lenses of this type were used extensively as projection lenses for many years. Figure 23 is a lens of this type.

A Petzval lens can form beautifully corrected images on a curved image which matches the Petzval curvature. These lenses are exceptionally well corrected for secondary spectrum (no conflict with the Petzval correction).

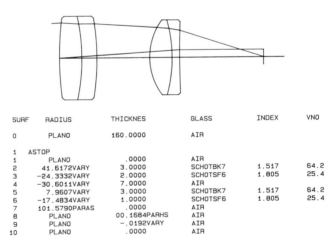

SURF	RADIUS	THICKNES	GLASS	INDEX	VNO
0	PLANO	160.0000	AIR		
1	ASTOP				
1	PLANO	.0000	AIR		
2	41.6172VARY	3.0000	SCHOTBK7	1.517	64.2
3	-24.3332VARY	2.0000	SCHOTSF6	1.805	25.4
4	-30.6011VARY	7.0000	AIR		
5	7.9607VARY	3.0000	SCHOTBK7	1.517	64.2
6	-17.4834VARY	1.0000	SCHOTSF6	1.805	25.4
7	101.5790PARAS	.0000	AIR		
8	PLANO	00.1684PARHS	AIR		
9	PLANO	-.0192VARY	AIR		
10	PLANO	.0000	AIR		

FIG. 23. The Petzval lens. $f = 16$ mm, $F/1.8$, half format 1 mm. (Designer: R. E. Hopkins.)

A 150-mm focal length Petzval lens covering a 15° half field with a curved field has as good an image at $F/1.5$ aperture as a 50-mm focal length lens used on a 35-mm camera operating at $F/3.5$.

3.4.8. Finite Conjugate Lenses

Most of the camera lenses are designed to image objects at infinity. If the lens is used at finite conjugates, it will not be corrected as well as the designer intended. The change in performance depends on the relative aperture of the lens and the extent of the magnification change. Some of the modern lenses are designed to operate at more than one conjugate position. This is done by using what is called a multiconfiguration optimization. Essentially the optimization is done for two or more conjugate distances simultaneously. This provides a design that works well at the different conjugates, however, it comes at the cost of reduced performance or more complexity.

The same design concepts apply to finite conjugate lenses as to infinite conjugates. A designer starting to design a finite conjugate lens will, depending on the magnification required, determine the total $F/$number requirement. For example, if the task is to design for a magnification of -0.25 (designers usually design from the long conjugate side to the short conjugate) and a numerical aperture (NA) of -0.20 on the short conjugate (the image side), then the NA on the object side will be 0.05. This calls for a lens with a working numerical aperture of 0.20 plus 0.05, which is 0.25, or an $F/$number of $F/2$. The designer would consider using one of the infinite conjugate lenses working at $F/2$ but would adjust the conjugates to work at a magnification of -0.25. The lens would have to be redesigned to operate properly at the finite conjugates. On the basis of symmetry, one can visualize how a finite conjugate lens must change to work at these finite conjugates. Take the double Gauss as an example. It is certainly capable of operating at $F/2$ out to at least a 20° field. If the requirement was to design for a magnification of -1, the lens should turn out to be reasonably symmetrical around the stop in the center. In order for it to work at a magnification of -0.25, one could expect that the lenses on the long conjugate should be scaled up to larger size and focal lengths, while the lenses on the short conjugate side should be reduced. The designer should then expect to allow the air spaces on the two sides of the stop to be variable. The design of finite conjugate lenses is essentially no different from designing an infinite conjugate lens, but it is necessary to realize that it requires a smaller $F/$number than the image-side numerical aperture requires.

Finite conjugate lenses are difficult to find on the shelf, for they are usually designed for a particular task; the variety is extensive but is not normally kept as inventory. When a finite lens is required, it is usually necessary to use an existing on-the-shelf lens which may or may not be designed for the conjugates required. It is then necessary to carefully test the lens at the conjugates needed to see if it performs adequately. The author has previously given warning about the need to be critical of its performance before the lens is accepted and imbedded in an instrument. As a rule, finite conjugate lenses have to be designed for the required task if the performance is important.

3.4.9. Summary of Design Concepts

The lenses described in this chapter represent the basic concepts used in lens design. The large number of designs available in the literature are variations of the concepts listed below.

1. Single elements always image on curved surfaces.
2. Elements of a single glass always have axial chromatic aberration.
3. By combining positive and negative lenses of different glass in the form of cemented or air-spaced doublets, they make an achromatic combination.
4. Usually all the elements in a system should be made achromatic. If they are not, the correction should be done close by to keep the colored rays together.
5. Positive elements can introduce positive astigmatism if the chief ray does not pass through their center.
6. The Petzval radius of a photographic-type lens should be between 2.5 and 8 times the focal length.
7. The Petzval curvature is controlled by the spacing and power of the positive and negative lenses. The spacing between the lenses may be air (see Fig. 14) or blocks of glass (see Fig. 17).
8. Closely spaced positive and negative lenses may be cemented together or allowed to have a difference of curvature. The latter case usually is called a critical air space, which is used to control high-order aberrations.
9. The strength of cemented surfaces can be altered by the choice of the dispersions of the glasses used. If the cemented surface is removed from the stop, it can be used to control the oblique spherical aberration. By placing the cemented surface to the rear of the lens as in the Tessar (Fig. 15), it will influence the upper oblique rays. The cemented surface in front of the stop will have a similar effect on the lower rays. The magnitude of the effect can be altered by the choice of the index difference across the boundary and by the strength of the cemented surface. The cemented surface

curvature can be changed by changing the dispersion difference between the two glasses of the cemented lens.

10. The rule for the correction of the Petzval radius is to use high-index, low-dispersion glasses in the positive lenses and the opposite in the negative lenses. An exception may be taken when designing telephoto or inverted telephoto lenses, where the first-order properties force a backward-curving field.

11. The correction of positive Petzval curvature and secondary color oppose each other. Expect to introduce secondary color whenever the Petzval curvature is made more positive.

12. Whenever large aberrations occur on a surface and are balanced out by following surfaces, the lens will be sensitive to centering, spacing, and conjugate change. Then, it is often prudent to add elements to reduce the sources of large aberration.

Modern methods of manufacturing of camera lenses have resulted in the elimination of several of the "no nos" by which the older lens designers were forced to abide. Today, extra elements can be justified as a trade-off for compactness and increased performance. Lenses may be made thinner if it helps to keep the size down. Unusual glasses may be justified if the manufacturing plant spends enough time to handle peculiarities which cause staining, chipping, or cracking during processing.

Designs for small production run numbers have to be made with considerably more conservative design rules.

3.4.10. Disclaimers and Words of Warning

The description of the lenses in the above sections was provided to illustrate the basic principles of lens design and to provide some insight and appreciation of how well lenses are corrected. The described lenses have a performance close to what one would expect to find in a manufactured lens of that particular type. However, one should not consider these as prescriptions for lenses to be made in an optical shop. A lens to be manufactured has to be subjected to a large number of manufacturing, mounting, and material considerations. For example, the prescribed glass may not be available in the small quantities needed or it may be too expensive. Curves have to be fitted to test glasses and tools if they are available. If they are not available, then new tools and test glasses have to be made. Test glasses can add as much as $400 per surface; new tools even more.

Some of the lenses shown above were obtained from patents. These have the additional problem that they seldom represent the lens as actually manufactured. The patent writers attempt to cover as much as possible, so they show several alternate designs and claim invention for all of them. Often the lenses are too thin, or the airspaces make it difficult to mount the lens. Sometimes the chosen glasses are not really available, etc.

Optical systems that are lens-limited (this means the imaging requirements are limited by the optical performance rather than the detector) should theoretically be reoptimized whenever the specifications are altered. If the F/number can be decreased, redesign can provide better performance. Relieving a tight tolerance on distortion will allow other performance parameters to improve. If a new design is to be undertaken, it is extremely important to challenge any specification that is inhibiting the performance of the lens. Trade-offs have to be made between a knowledgeable designer and the user.

3.4.11. Aspheric Surfaces in Camera Lenses

Aspheric surfaces intentionally designed for glass lenses are seldom used, for even today, the cost of making a precise aspheric surface is many times the cost of a spherical surface It is usually more economical to use extra lenses with spherical surfaces. The exception to the rule is when some of the lenses are to be made in plastic. If the production numbers are large enough to cover the high cost of the molds, aspheric surfaces are used. Aspheric surfaces become much more useful in large optical systems using mirrors, but precision aspherics are extremely expensive, requiring specialized equipment which limits the numbers that can be made. A positive all-spherical lens separated from a negative lens can be designed to provide much of the same correction as an aspheric corrector plate.

The Kodak disc camera uses an aspheric lens made by molding glass. This represented a significant technological advance, but the process is expensive and requires large production to justify the tooling costs. The lenses also have to be small in diameter.

The effort that has gone into making aspheric surfaces for production lenses exceeds the rewards from using them. Now that coated lenses are so well developed, it appears that the time for aspheric lenses in cameras has not yet arrived. Extra-spherical lenses appear to be the way to go for production runs. Some of the most modern diamond-grinding machines are beginning to be capable of grinding precision aspheric surfaces which are smooth enough to polish without altering the surface contour. For special ultraprecise lenses, these machines probably can make it feasible to use several aspherics in a single design and reduce the number of elements. This, however, will not come about until an application of sufficient value can be found to finance the development of the machines and the technique.

It is well to remember that the primary need in lenses is to find a way to correct the Petzval field curvature. Making the lens achromatic is also a major portion of the design task. Aspheric surfaces do not help to provide the necessary field flattening and they do not help in the color correction. Aspheric surfaces merely help to reduce residual aberrations. If the aspheric is placed near the stop, it primarily helps to reduce the high-order terms of spherical aberration. When the aspheric surface is located away from the stop, it effects the top or the bottom of the beam and introduces all the high-order aberrations. It is unlikely that it can balance all of the high-order aberrations from the spherical components.

3.4.12. Camera Lenses for Use with TV Cameras and CCD Arrays

Most of the TV lenses are of the same types as camera lenses for photography. They usually have to be color-corrected to match the spectral sensitivity of the TV cathode. The image tubes seldom have as good a resolution as photographic film. Thirty lines per mm is a representative limiting resolution. Since the TV camera cannot use a resolution higher than this, the users want the lenses to have as high a response at frequencies below 30 lines/mm as possible.

Some of the recent TV systems are using CCD arrays with elements in the 10 to 15-μm range. Optimized designs should concentrate as much of the total energy as possible into pixels the size of the detector pixels. On the other hand the images should not be so small that they can land in between the detector units. This would result in loss of information. The image pixel should cover a sizeable portion of the two adjacent detectors when imaged between them. Aside from the discrete nature of the detectors and their different spectral responses, the TV lenses are similar to photographic lenses.

3.4.13. The Vast World of Zoom Lenses

Zoom lenses have been described[4] since 1930, but the real explosion in their use came about after 1950. There were several reasons for the delay in their widespread use. First, they represented a huge computational task before the days of electronic computers. The theory was reasonably understood but the design problem was overwhelming. The widespread use of TV provided a major requirement for zoom lenses. Eight-millimeter movie cameras were the first to use zoom lenses, because the early models were too big and cumbersome for larger frame sizes. Finally, the power of the computer and advances in the manufacturing of lenses with reduced thicknesses lead to the compact zoom lenses for 35-mm cameras and compact lenses for TV. There have probably been more patents for zoom lenses issued within the last ten years than for all other lens systems combined.

There are two fundamentally different zoom lenses.[5,6] The earliest models were called optical zoom lenses. In these lenses, alternating lenses are moved together. As the lenses slide along the optical axis, the back focal distance changes, but there are positions where the back focal lengths are at the same distance from fixed elements. It has been shown that there are as many stationary points as there are movable spaces.[7] For example, if there are three elements, as in the standard triplet lens, and the two positive lenses are tied together to move axially with respect to the negative lens, then there are three movable spaces. When the powers of the lenses are chosen correctly,[7] there will be three positions for the sliding lenses, where the distance from the negative lens to a focal plane is the same. The moving spaces are the distance from the first positive lens to the negative lens, from the negative lens to the second positive lens, and from the second positive lens to the focal plane. This is true when the object is at infinity. When the object is at a finite distance, there are four movable spaces, so it is possible to have four stationary points. The amount of focus error between the stationary points depends on the range of focal-length change.

Another requirement for a zoom lens is that the numerical aperture of the imaging beam must be stationary as the focal length changes. Usually the movable components are in front of a stationary focusing lens. The zoom components also operate at low numerical aperture in order to keep the aberrations small. The rear focusing lens is used to increase the overall numerical aperture of the lens. A zoom change in focal length of three is about the limit that one can expect from a single optical zoom lens. To obtain larger zoom ratios, it is necessary to use tandem zooms.

Mechanical zoom systems move the positive and negative lenses by using cams. These can continuously vary the focal length and maintain a stationary back focus. Mechanical zoom lenses were out of favor for many years because the cams were expensive and subject to wear. Today, the mechanical zoom has essentially overtaken the optical zoom. This has come about because numerically controlled lathes and new materials have helped overcome the wear problem. One aspect of the design of a zoom lens is to use designs which do not require rapid changes in the rate of change of distances between the lenses. Sudden changes in the cam profile can be a source of wear.

Zoom lenses for movie cameras have to have a stationary image plane during the zoom, for the focus must not change from frame to frame as the lens is being zoomed. Still cameras, however, do not have this requirement, because the lens can be focused for each position of zoom. This has made their design easier and has made them practical in still-camera photography.

CAMERA LENSES 97

The author designed several optical zoom systems in early 1950, just at the time when digital computers were being developed. The zoom lenses were designed using the IBM card program calculator. This marvelous device was such an improvement over the Marchand mechanical calculator that we drove 90 miles to use it. It had a storage capability of 10 numbers of 10 digits. It could trace meridional rays at the blinding speed of 40 seconds per surface. The PC computers of today can trace 400 surfaces per second. The Control Data 7600 machines can trace 1000 surfaces per second. In 1950, there were no available optimizing programs that could handle the ray tracing. The optimization had to be done with third-order optimization and then be reset for new target values after if was ray-traced. This experience along with the computing bills probably set up a negative prejudice towards zoom lenses, and so the author has had little experience in the use of modern zoom-lens design. This story has been told to warn the reader that the following statements about zoom lenses may be coming from a prejudiced viewpoint.

Today, zoom lenses are extremely popular, and since they are being made in large production runs, they are available at reasonable prices. It should be remembered, however, that zoom lenses do not have images as good as a lens designed for a fixed set of conjugates. The zooming feature requires many more elements in order to obtain good imagery. A typical $F/2$ single-focal-length lens may have six elements, while a 3 times zoom lens will require 12 to 15 lenses and will be much longer and larger in diameter. These extra lenses and the internal movement of the lenses are sources for image degradation. If a new application appears to require a zoom lens, the following considerations should be reviewed.

If a standard zoom lens can be found to meet the requirements, then use it.

If the requirement is lens-limited, do not consider using a zoom lens. It would be better to use fixed lenses on a turrent. There are few cases where the continuous focal change of the zoom lens are required. In fact, most zoom lenses are used at the two extremes of their focal lengths. The zoom lenses on 35-mm cameras are used primarily as an aid to framing the scene.

3.4.14. Mirror Systems

Mirror systems lead one into the world of the astronomers, for they have been the primary users of mirror systems to form images. They were forced to use mirrors for two reasons. First, to overcome the secondary spectrum which is inevitable with long-focal-length, large-aperture systems. Second, the large apertures required for astronomy ruled out the use of large glass lenses. Most mirror systems have small fields and long focal lengths. The aberration correction has to be done by using more than one mirror, and the principles are essentially the same as with refractive lenses. In order to correct Petzval curvature, collective and dispersive surfaces have to be used at different apertures. Coma and astigmatism are controlled by introducing spherical aberration at positions in the system which have differing values of the height of the chief ray compared to the axial ray ($\bar{y}/y$). In order to correct coma, it is necessary to have more than one source of spherical aberration of opposite sign. The Schmidt telescope is coma-free because the chief ray is normal to the mirror.

Mirror systems almost always require aspheric surfaces: conic sections are commonly used. The large sizes of telescopes allow the use of local grinding and polishing; although not easy, it can be done and is a reasonable part of the cost of the telescope. Large telescopes are not made on production lines.

There has been so much written about large telescopes that this chapter is not the place to say more about them.

Mirror systems are not competitive with glass lenses when the diameters of the elements are 8 inches or less. Refractive elements are usually preferred to mirror systems, provided the glasses can transmit the required wavelengths. Mirrors do not have optical invariants as large as refracting systems. There are always problems with obscuration with mirrors. Many of the smaller telescopes used as field scopes appear to have less chromatic aberration than an equivalent refractor, but their resolving power is not up to the refractor. The main advantage of small reflecting telescopes is the light weight and short length that can be achieved.

Conic surfaces are mentioned so often in optics text books that they are often misused. For example, the parabola can focus a beam from a distance point source to a perfect focus, provided it is focused on the single optical axis of the parabola. Many times, the source is not located on the axis, so the image formed has coma and may be worse than one would obtain with a sphere. The same is true for an ellipse. It too has a single axis for the object and image foci. If the source is not on this axis, the image will have a large amount of coma. The great virtue of a sphere is that it has one center of curvature and an infinite number of axes.

The following reminders are offered to help in making the decision to use mirrors or refractors.

It is usually better to use refracting systems if the diameters of the lenses are less than 8 inches. The only advantages of mirrors are for cases with a wide spectral range or one beyond the transmission range of available glasses or workable materials. Mirror systems sometimes help to make a system more compact, but usually at the cost of obscuration and baffle problems.

Avoid using mirrors to reflect an imaging beam around corners while focusing the beam. The curvature on the mirrors introduces large amounts of astigmatism and coma, which are hard to correct. This is practical if it is merely a light-gathering problem with no resolution required. Some of the solar collectors do this effectively.

The two surfaces of a single lens can be polished and cause less scattering than a mirror that has to be coated for high reflectivity. Even if the lens is coated for antireflection, it will scatter less than a mirror.

A plane mirror placed in an optical system has to be flatter than often realized. Selected flat ground plate glass used to be used in optical systems, but when float glass put most of the plate glass companies out of business, the only source was float glass. Often float glass is not flat enough for optical performance. Just as many new instruments were being developed, the procuring of high-quality flat glass became a problem. Fortunately, planetary polishing machines have been highly developed, and now high-quality large flats can be obtained for reasonable prices.

3.5. Projection Lenses and Condensers

Projection lenses have a great deal in common with camera lenses, but there are differences. Camera lenses are seldom suitable for projection. Projection lenses tend to have longer focal lengths and narrower field angles. The relative apertures range from around $F/3.5$ to $F/1.5$, but they seldom are pushed to the limits of speed found in modern camera lenses. There are not as many projection lenses produced as camera lenses, so they cost much more. This is probably why projection lenses usually are not corrected as well as camera lenses. The importance of their performance is not considered to be as high a priority. A person will pay several hundred dollars for a superb camera lens and then be content with a $20 projection lens.

A condenser lens and projection lens should be designed to match each other. Many a good projection lens has been blamed for faults in the condenser. In the following sections, two basic projection systems will be discussed.

3.5.1. Koehler Illumination

The most widely used system for projecting transparencies is the Koehler system which is illustrated in Fig. 24. In this schematic drawing, the source is assumed to be a disc of light source which is Lambertian. This means

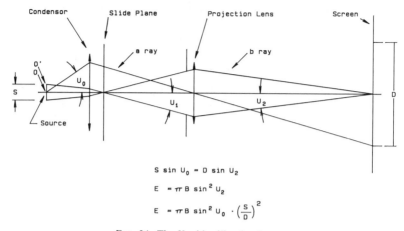

FIG. 24. The Koehler illumination system.

that the surface of the source of light has the same brightness when viewed from any angle. In modern terminology, the word brightness should be replaced with luminance. In some ways, this is a shame, for many people seem to have an understanding of the difference between brightness and illumination but have trouble between luminance and illuminance. Luminance and illuminance are appropriate terms when writing about visible light. Illuminance is the modern replacement for illumination. Radiance and irradiance are generally equivalent when discussing other than visible radiation. However, when the subject matter involves physical optics topics, such as interference and diffraction, the term intensity is commonly used in place of irradiance. Intensity is used to represent the square of the amplitude of a wave front. Irradiance, illuminance, and intensity all refer to flux (photons/sec/unit area). This chapter will use intensity in the section on laser optics.

The diagram shows the two rays normally traced through an optical system. The a ray starts at the center of the light source and is imaged to the center of the projection system and then passes on to the projection screen. The b ray passes through the center of the slide plane, expands to the projection lens, and is focused on the projection screen. The optical invariant equation [Chapter 2, Eq. (2.27)] states that:

$$S \sin U_0 = D \sin U_2. \tag{3.7}$$

The illuminance on the screen is given by the equation

$$E = \pi B \sin^2 U_2, \tag{3.8}$$

where B is the luminance of the source (lumens/solid angle/cm). The illuminance can be written

$$E = \pi B \left(\frac{S}{D}\right)^2 \sin^2 U_0. \quad (3.9)$$

The last equation shows that the illuminance on the screen depends on the source luminance, the square of U_0 (referred to as the collection angle of the condenser), and the ratio of the source diameter to the screen diameter. The illuminance at any point on the screen will be determined by what is seen by an observer standing in front of the screen and looking back at the projection lens. The observer will see the light source filling the aperture of the projection lens. The angle of subtense of this image of the source and the source luminance will determine the screen illuminance at the position of the observer.

The real situation is not quite as simple. Light sources are seldom flat discs with Lambertian luminance. The closest to it is a ribbon-filament tungsten lamp. It does not have the surface luminance of a closely-packed-coil on coil filament in a projection lamp. Many projection lamps use coil tungsten filaments. This filament has varying degrees of luminance and is not a continuous source. The precise calculation of the screen illuminance then is an extremely difficult calculation to make. It is necessary to trace rays from every point on the screen to a grid of points on the exit pupil of the projection lens and extend them back to the source. If they hit a part of the filament, then the luminance in that direction of incidence has to be used. Detailed knowledge of the light source is not known to this degree, so one can really only make estimates of the uniformity of the screen. For most projection systems, uniformity of 10% is considered to be adequate. Any tighter tolerance should be considered with caution, for usually the source information is not known well enough to make a meaningful calculation, even if done with all its complexity. It is important to be able to rely on experience with similar systems.

The projection system will always require a trade-off between the ratio of S/D and the angle U_0. The burden is usually placed on the collection angle of the condenser, and it may increase to 35 or 45°. It is not wise to extend the collection angle beyond 45°. This requires too large a ray bending for the condenser lenses. The spherical and chromatic aberrations become difficult to control. Aberrations in the condenser are often considered unimportant. If the source is large enough and is Lambertian, then rays from all points on the screen will pass through the exit pupil of the projection lens and eventually hit the source. In this unrealistic case, the aberrations of the condenser do not matter. When sources are not Lambertian and when it is not practical to use a source large enough to supply the aberrated rays, the screen will not be uniformly illuminated.

The remaining trade-offs are the size of the screen and the source luminance. The net result is that projection systems require uniform high-luminance sources. Tungsten, carbon arcs, and high-pressure gas sources are the choices today. The laser provides the luminance but not the source size, so it does not help in the conventional projection systems.

When tungsten sources are used, a mirror behind the coils can be adjusted to image the coils back between the coils and thus make the source to have more apparent uniformity. Unfortunately, the mirror has to be large enough in diameter to form a reasonable image. It has to be accurately positioned. Most household projection systems have these mirrors, but they seldom do the slightest bit of good, for they are too far out of adjustment to put more light on the screen. Tungsten projection lamps do a reasonably good job of providing a uniform light source, but they are deficient in blue light for some applications. The high-pressure mercury sources with additives are commonly used for broader ultraviolet requirements. They have high illuminance which is far from being uniform. There are one or two hot spots near the poles of the arc. In order to use these lamps, the designer of the condenser system must decide how much of the luminous area of the lamp to use to fill the aperture of the projection lens. As with the tungsten source, the arc lamp is viewed from different directions as the observer moves from one part of the screen to another. The luminance changes for different viewing angles. The lamp manufacturers provide polar diagrams of the radiance of their light sources. In order to calculate the irradiance in the image plane of a projection system it is necessary to know the point to point source radiance, as viewed, from all the angles for the rays passing through the condenser lenses. This information is almost never accurately known, so the calculations may be subject to errors in uniformity of as much as 10%.

There is one outstanding application which requires extremely high uniformity of illumination. It is in the projection systems used to print wafers in the manufacture of microcircuit chips. The industry is now beginning to ask for 1% uniformity. This requirement has led to several new systems for obtaining uniformity without overpowering the problem with large light sources. Most of these systems use some form of scrambling of the light. Ground and opal glass can do a good job of scrambling the light, but it causes severe loss of source luminance.

Designers have used scrambled fiber bundles. These scramble the light uniformly, but the luminance suffers. Numerous segmented mirrors and fly's-eye lens systems have been proposed, but they are usually too expensive to make. One of the most popular systems uses a light-integrating bar. Figure 25 is an illustration which will be used to explain how they work. The short arc lamp is imaged on one end of the integrating bar. This can

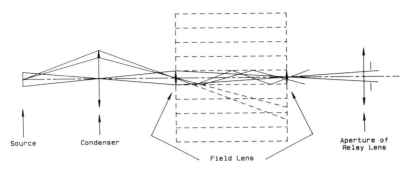

FIG. 25. A light-integrating bar.

be done with a lens or mirror condenser. A field lens is placed over the end of the bar, which images the aperture of the condenser on to the exit end of the bar. The bar has a square or hexagonal cross section. The walls of the bar are polished to optically flat surfaces. The integrator is like a kaleidoscope. It does not completely integrate but splits the light up into nonuniform beams which undergo odd and even numbers of reflection. At the other end of the bar, the odd and even reflections have flipped the uniformities over so that they tend to make the distribution more uniform. The square and the hexagonal cross sections allow tight packing of the entrance end of the rod as seen from the other end. The hexagonal cross section is more expensive to make, but it does provide more integration. Higher numbers of sides would not allow complete packing, so there would be gaps in the source as seen from the exit end.

The exit end of the bar can now be used as a new source which is more uniform. This may be imaged with magnification onto the mask plane. The mask plane can have a field lens which images the magnifying relay lens on the aperture of the projection lens. The illuminance at the end of the integrating bar is uniform because the total flux from the multiple reflected images emerges from the end, but as the light propagates, it becomes nonuniform in a pattern similar to the multiple images of the entrance end of the bar. The illuminance on the relay lens will have this pattern and will be relayed to the projection lens aperture. This puts an apodizing amplitude across the aperture, which in theory has an effect on the diffraction image formed on the wafer. This however, is a small effect if the nonuniformity on the entrance aperture of the bar is not too great and the integrating bar's entrance face is completely filled. A light source which is not uniform and Lambertian can, with specular systems, be made spatially uniform in discrete planes. In between, the illuminance will not have uniform illuminance. It is necessary to completely diffuse the light, as can be done with an integrating sphere, in order to have uniformity throughout the optical system. Complete diffusion, however, lowers the source illuminance, which is usually too heavy a price to pay.

3.5.2. Summary of the Method for the Determination of Uniformity of Illuminance

In order to calculate the illuminance on a screen resulting from a projection system, it is necessary to take a point on the screen and trace rays back through the exit pupil and all the way to the light source. The rays should be traced through the centers of a uniform grid pattern on the exit pupil of the projection lens. (The grid should actually be located on the aperture stop of the projection lens, and the grid on the exit pupil may be somewhat distorted by the lenses between the aperture stop and the screen. This effect, however, usually is neglected because it is small if the projection angle of the projection lens is not large.) The rays must pass through the entire optical system to the light source. A ray that misses any part of the source must not be added to the summation. The luminance of the source in the direction of the ray should be used in the calculation. Unfortunately, this information is not usually available, for lamp manufacturers usually measure the total flux from the source in a given direction.

When the projection angle is large, it is necessary to take account of the cosine law for the angle of incidence on the screen. The exit pupil may also be distorted and will not subtend the same solid angle as it does when viewed from the center of the screen. Some of the latest projection optics for microcircuit lithography are designed to be telecentric on both sides of the lens, partially in order to eliminate any cosine losses. All these complications are offered as a warning to never underestimate the difficulty to provide uniform illuminance over an extended area.

3.5.3. Partial Coherence Effects in Projection Systems

The above discussion on uniformity of illuminance leads to still another detail that a designer must consider when designing precision projection systems. This is the effect of partial coherence in optical systems. There are many discussions of this subject in the literature, including Chapter 4 in this book. A designer, however, has to have a way of understanding the problem in order to help decide what to do about it. This explains the following discussion.

Sources like tungsten or short arcs are incoherent sources. They can be considered as an assembly of completely independent sources of coherent light. The phases of the light from these independent sources are unrelated and so will not interfere with each other. The condenser lens images the source points onto the aperture of the projection lens (see Fig. 24). The size of the independent source points are determined by the NA of the condenser. The diameters of the points O and O' are given by the equation (the Van Cittert theorem)

$$d = 1.22 \frac{\lambda}{U_0}. \qquad (3.10)$$

The light from each point on the source forms a single-phase wave front which is brought to focus into the aperture of the projection lens and forms Airy disc images if the condenser lens is aberration-free (it seldom is). Now consider the transparency in the slide plane as being a fine grating with lines running perpendicular to the plane of the paper. The grating will transform the wave front from the point O to zero order and plus and minus first- and high-order diffraction images in the plane of the projection lens aperture. The number of orders will depend on the grating frequency and the diameter of the projection lens aperture. The same thing will happen to the wave front from the point O', but now the zero order is located at the edge of the aperture of the projection lens. The high orders of diffracted light get through for one side of the pattern but are cut off on the other side by the projection lens aperture. This means that the light from the point O on the source will form a more faithful image of the slide than the light from O', because the central image passes the high-order frequencies symmetrically. On the other hand, the light from point O' passes higher frequencies on one side, so it is providing some higher-frequency information. In order to calculate the modulation in the screen, it is necessary to calculate the orders of diffraction that pass through to the screen plane for each independent point on the source. The final image is then constructed by adding the intensities from each point, because they are incoherent.

There is definitely a trade-off to be made. When the source completely fills the aperture of the projection lens, some high-order diffraction contributes to the image. If the source is imaged to underfill the aperture of the projection lens, some of the higher-order information is lost, but the plus and minus orders of diffraction are more symmetrically passed. It is difficult to make any general statement as to what is best. The precise calculation is an order of magnitude more extensive than is required to calculate the optical transfer function (OTF) for an incoherent source. Calculations have to be made on the effect of partially filling the aperture for specialized objects. They show that edges may be sharpened at the 50% level of intensity, but at the lower intensities it is broadened, and at the higher intensities there are unreal fluctuations in intensities. This is useful for cases where the recording medium has a narrow threshold and the targets are specialized. Under most conditions, it is of dubious value to calculate the partial coherence. One has to face the possibility of seeing artifacts whenever viewing close to the classical limits of resolution. Past experience has shown however, that better viewing is obtained by partially filling the aperture. The optimum value is around 80% of the aperture. This is partly because it keeps the main intensity of the beam passing through the projection lens away from the retaining cells that hold the lenses. These become sources of scattered light which can dilute the delicate detail which one is trying to observe.

3.5.4. Abbe Illumination Systems

When the aperture of the transparency is smaller than the aperture of the projection lens, the source can be imaged directly onto the transparency, if the source is uniform. This is done in movie projection systems and in microscope systems. They do not use a field lens in the transparency plane, so the condenser lens is not imaged sharply in the aperture of the projection lens. With this kind of illumination, the projection lens should also not be completely filled, and the effects of partial coherence are similar.

3.6. Lens Systems for Laser Systems

Most of the optical lens systems used for laser systems are either designed for collimating beams or for focusing collimated beams. These two applications are reciprocal. A good focusing lens will work equally well as a collimator. The collimating lenses can be of lower numerical aperture than the focusing lenses, because focusing lenses are usually used to form small spot diameters of light. As far as design or use is concerned, the two applications are the same. Laser applications usually require that the converging wave front to an image point be spherical to within a fraction of a wave. One-tenth of a wave (peak to valley) is a common specification. Much of the value of the high luminance of the laser is lost if the wave-front deviations exceed the 1/10 to 1/4 wavelength limit.

Most laser beams coming directly out of the laser have a Gaussian profile of intensity across the aperture. This has lead to considerable confusion in lens specifications and calculations of the spot sizes encountered at the focus of a laser beam. Prior to the laser, optics people used the Airy disc image diameter as a reference. The usual description of the image of a perfect wave front assumed that the intensity of the light entering the lens was uniform across the aperture. The diameter of the first dark ring in the pattern could be easily measured, so it was used as a measure of performance.

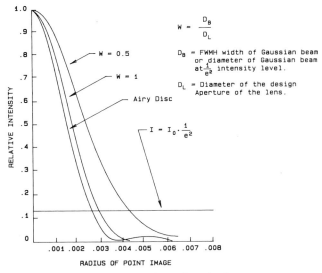

FIG. 26. Focusing a Gaussian beam.

The diameter of the ring is

$$d = 2.44 \lambda F/\text{number}. \quad (3.11)$$

Gaussian beams which are not truncated do not form the diffraction rings seen in the Airy disc. Since the Gaussian beam theoretically tapers off to zero intensity at infinite aperture, any real optical system truncates the Gaussian beam. A common procedure is to state that the Gaussian beam is truncated to the $(1/e)^2$ intensity level. This means that the lens is doing the truncating. The diameter of the image measured at the $(1/e)^2$ intensity level is given by:

$$d = 1.83 \lambda F/\text{number}. \quad (3.12)$$

The number 1.83 is less than 2.44, so a common belief is that the Gaussian beam is smaller than the Airy disc. This is misleading because the Airy disc diameter indicates the radius of the first dark ring, while the Gaussian diameter represents the 13.5% intensity level. The Airy image will actually concentrate the light into a sharper image than the Gaussian beam. This is illustrated in Fig. 26. The intensity profile of the Airy disc is more compact than the Gaussian beam. The sharpness of the intensity profile is increased by allowing rays at larger diameter to enter the lens. The limiting resolution is proportional to the numerical aperture of the largest aperture. If the aperture of the lens is increased, it is true that the Gaussian beam will become sharper and provide more resolution, but the numerical aperture of the lens will be increased, and the aberrations must be corrected for the larger aperture. The same is true for the uniform intensity across the wave front.

The final statement then is that the uniform intensity forms the sharpest image of a laser beam. To obtain the uniform intensity across the aperture of the lens, the laser beam must overfill the aperture, so there is a substantial loss of light. The optimum compromise between loss of light and maximum resolution is to truncate the laser beam with the aperture of the lens as the $(1/e)^2$ intensity point.[11]

There is another confusion that has cropped up in the Gaussian-beam literature. It is the question of the position of the waist in a Gaussian beam. It is pointed out by many authors that the waist in a focussed beam is not located at the paraxial focus of the perfect lens. This is true and is particularly noticeable when the numerical aperture of the lens is small. On the other hand, the sharpest image one can obtain is located at the paraxial focus. The waist lies towards the lens, but the intensity of the light at the center of the image is not as high (the Strehl ratio is less) than it is at the paraxial focus. The reason is that when the perfect converging wave front is formed at the paraxial focus, the light from the entire aperture arrives at the center of the image in phase. At the waist, some of the wave front is out of phase at the center of the image. The image at the waist is slightly smaller than the image at the paraxial focus, but it does not have as good a Strehl ratio.

There is one more practical bit of information about the use of laser beams. Laser optics usually have to form diffraction-limited images. This means that anyone procuring a focusing or collimating lens should test the lens to make sure it performs as advertised. To obtain a quantitative check, one has a difficult measurement task in measuring the spot diameter when using a Gaussian beam. To check the above formula for spot size of the Gaussian beam, it is necessary to measure the beam at the $(1/e)^2$ intensity level. There are instruments available to measure this, but they are not cheap, and there are many ways one can be fooled. For example, the optics may not be clean when the measurements are made. Nonuniformities on the sensitivity of the detector may give erroneous results, for the Gaussian beam is not uniform. The simplest, most economical way to check a diffraction-limited lens is to use a uniform intensity beam and check the Airy disc image visually. The diameter of the first dark ring can be measured with a filar micrometer in a microscope; or the image may be projected on a wall, using a microscope objective, and be measured with a ruler. If there is any residual aberration, the image will not be symmetrical in appearance as the image is observed in front and in back of the best focus. This is an extremely sensitive test. Errors less than 1/10 of a wave can readily be detected. Any coma in the image will change the uniformity of the intensity in the first bright ring. Small amounts of astigmatism will form fourfold accumulations of intensity in the bright ring. The author will not accept a lens specified to require a 1/10-wave performance without looking at the Airy disc diffraction patterns. Lenses of this performance are costly, and not many shops have the equipment to measure it. It takes an interferometer in the $100,000.00 range to do it. The other side of the coin, however, is that the lens should not be specified to this tolerance if the requirement does not demand it.

The simplest collimating lens is a single lens. It can be bent into a shape which will minimize the spherical aberration and have no coma. The absence of coma increases the tolerance on how accurately the image has to be formed on the optical axis of the lens. Since the spherical aberration can only be minimized with the shaping of the lens, there is a limiting aperture for which the lens can be corrected to be diffraction-limited. Figure 27 is a plot of the OPD versus the F/number for single lenses of BK7 glass bent into the zero-coma position. The plot shows how the OPD varies with wavelength and beam diameter. There is an effect of change of wavelength and the glass used for the lens, but the curves provide help in deciding whether a single lens has a chance of meeting the requirements on OPD. The zero-coma condition shapes the lens to have a strong curvature on the long conjugate side and weak curvature on the short conjugate. In the case of BK7 glass, the zero-coma position is close to a plano-convex lens. Plano-convex lenses can be found directly in the catalogues of well-known

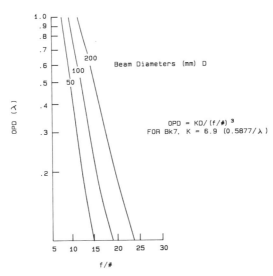

FIG. 27. Plot of OPD versus F/number for single lenses of BK7 glass with zero coma.

optical-component suppliers. Lenses made of higher index of refraction will have smaller spherical aberration than BK7 glass.

When the application requires better correction than the single lens can provide, the next alternative is a two-element lens. Normally, doublets are corrected for chromatic aberration as well as spherical and coma. A lens to be used with a single wavelength, such as 0.6328 μm, does not need the chromatic correction. It takes a positive and negative lens to completely correct for the spherical and coma. However, two positive lenses can reduce the spherical aberration from that of a single one. This is done by using a weaker-front single lens which is bent into the minimum-coma position, and then by adding a positive lens which is bent into what is called the aplanatic position. An aplanatic lens is shown in Fig. 28. The rear element receives a convergent beam from the front lens. The convex surface can be adjusted to satisfy the aplanatic condition and introduce no spherical aberration or coma. The aplanatic condition is satisfied when the radius is adjusted so that $\sin U'$ is equal to the index of refraction of the glass times $\sin U$. The last surface is concentric with the convergent wave front and so it introduces no aberration.

Doublets using a positive and negative lens corrected for laser beams usually use two glasses of the same high-index glass. This, however, requires that the lenses be air-spaced. Some optical houses prefer to make cemented lenses, but this requires different glass types. Stock lenses are not available which are designed for laser applications, and the wave front distortions are guaranteed to be within 1/4 to 1/10 wavelengths. They are expensive, for they are difficult to make to this quality, and the production numbers are relatively small. There are surplus-lens suppliers who buy obsolete military optics. Sometimes it is possible to find a lens corrected to these

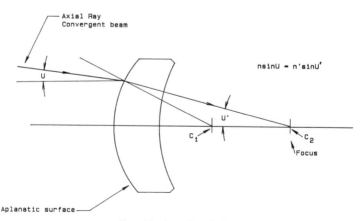

Fig. 28. An aplanatic lens.

tolerances at ridiculously low prices, so it is worth investigating. Many surplus lenses are not suitable, which one can tell by looking at the two outside curves of the lens. A good collimating lens will have a short radius for the long conjugate side and a slightly convex surface on the short conjugate side. Doublets with an air space can be designed to work up to $F/3.5$ for apertures of 50 mm and be diffraction-limited.

Beam diameters above 50 mm probably require three elements. The extra lens should be added to the short conjugate side of a well-corrected doublet. It should be bent into the aplanatic form. The last surface is then concentric to the final image point, so it adds no aberration. If the three-element lens is designed as a unit, the radii would deviate slightly in order to avoid the concentric last surface. A surface that reflects the light directly back on itself is usually undesirable for laser application. By adding still further aplanatic lenses to the lens, the numerical aperture can be increased still further. Lenses used for collimating light from laser diodes have numerical apertures as high as 0.8. With numerical apertures this large, the working distance and depth of focus are extremely small. Often diodes are covered with thin coverslips of glass. The thickness of the glass must be taken account of in the design when such large relative apertures are being used. A thin plate of glass introduces positive spherical aberration when placed in a converging wave front.

Laser-focusing lenses are usually not corrected for astigmatism or Petzval curvature. The field coverage is therefore limited; consequently, the diode must be carefully located on the optical axis of the collimating lens, or the emerging wave front will not be collimated with the expected quality.

The author made a study of the field coverage of laser-focusing lenses. The results are given in Table IX. This chart shows the field angles and the number of independent diffraction image points that single and double lens laser-focusing lenses can image. The spot diameters provided are the

TABLE IX. The Performance of Single and Doublet Focusing/Collimator Objectives. The Results Show the Price Paid for Using More Compact Optical Systems

Lens type	Beam diam. (mm)	$F/\#$	Field diam. (mm)	Spot diam. (mm)	N (spots)	Field diag. (deg.)
Single	50	20	28	0.031	815,000	0.8
Single	100	20	38	0.031	1,500,000	0.5
Doublet Aplanat	100	11	16	0.017	886,000	0.42
Doublet + −	100	5	5	0.008	422,000	0.29
Doublet + −	100	2.5	2.5	0.004	411,000	0.09

diameters of the Airy discs. As the F/number of the lens is reduced from $F/20$ to $F/2.5$, the number of spots and the field-angle coverage is steadily reduced. It was surprising that the single lens was the winner for its ability to transport information. The single lens, however, requires a long focal length. It uses a lot of space, and the air path will cause some degradation in the images. The higher-speed lenses shorten up the air path and can be packaged in a smaller volume. This is the price of miniaturization. Larger numerical apertures reduce the field. Any attempt to expand the field will increase the complexity of the lens or reduce its performance.

There are laser lenses that require good imagery at several wavelengths. If the requirement will allow a focus change, the lenses can be designed with variable spacing to provide excellent images. However, if the lens must work at two or more wavelengths, then the lens must be fully achromatized. This can result in considerable loss of performance or increased complexity of the design. As mentioned in Section 2.3.5, normal achromats can focus only two wavelengths at a time to a common focus. If the wavelengths are too far separated, there will be considerable variation of spherical aberration with the wavelength change. This can be reduced by using airspaces or lens thicknesses, but it adds to the complexity. When dealing with diffraction-limited lenses, it is important to keep complexity down. It is not a trivial problem to make and mount lenses while maintaining diffraction-quality wave fronts. The lens becomes a sensitive interferometer.

3.6.1. Lenses for Laser Scanning

Scanning is a major use for laser lenses. Scanners are used to print and read information at high speeds. Laser scanners are finally beginning to compete with the cathode ray tube for displaying documents and images. There are some interesting design problems involved in scanners. A user has to consider many attributes of a prospective system. Some references are provided.[8] This section will concentrate on a discussion of the lenses.

Most of the systems use prelens scanning instead of postlens scanning. To do the scanning, postlens scanners use a rotating mirror between the lens and the image. This method has the advantage that the lenses can be simple collimating lenses as described in the previous section. This system scans onto a curved surface, centered on the axis of rotation of the mirror. If the application can accept the curved scan plane, it is the preferred way. Prelens scanning requires that the rotating mirror of the scanning device be located in front of the lens. This mode of scanning is chosen when the curved scanning surface is not acceptable. For prelens scanning, the lens must be able to form images on a flat image plane.

Prescanning requires a lens with an entrance pupil in front of the lens. This makes the design similar to an eyepiece of a telescope. All the optics are on one side of the pupil. It is difficult to correct all the third-order aberrations. Distortion and lateral color are particularly difficult to correct. Fortunately, most of the laser-scanning systems use monochromatic light, so the lateral color is not a problem. Laser-scanning lenses are often designed to have negative distortion, so that the image height is proportional to the field angle instead of the tangent of the field angle. Due to several other factors, laser-scanning lenses seem to cooperate with the designer and scanning lenses can be designed with remarkable results even when they are basically unbalanced systems.

There are three systems used for scanning, namely, polygon, galvanometer mirror, and hologon scanners. Each of these scanners imposes slightly different constraints on the optics. There are many recent papers written on the advantages and disadvantages of using these scanners.[8-12] A few of the differing requirements for these systems will be mentioned, but there are many trade-offs to be made. Anyone contemplating a scanner design must collect a file on this subject.

3.6.2. Galvanometer Mirror and Polygon Scanners

These scanners are particularly popular with electrical engineers because the mechanical and the optical requirements are simplified. Usually these scanners do not require a strict meeting of the F-theta condition. The rotation of the mirror can be controlled electronically to adjust the image height. Galvanometer scanners cannot scan at speeds as high as the polygon or holographic scanners. Some scanners scan in two dimensions. These usually use polygon or hologon scanners for the fast scan and a galvanometer mirror for the slow scan. These hybrid scanners can cause some strange distortions which must be corrected optically or with the electronics. Galvanometer systems are sensitive to vibration and may not be suitable for instruments that have to be subjected to vibration shocks.

Polygon scanners can rotate at high speeds and scan faster than galvanometers. However, they are expensive to make. Usually polygon scanners call for lenses that follow the F-theta law. One of the difficult problems in making a polygon scanner is to have all the faces parallel with the rotating axis. The polygon should have no, or only small, pyramid errors. There should also be no wobble in the axis of rotation. These two errors cause what is called cross scan errors. Many attempts have been made to correct these errors by using cylindrical lenses to focus the collimated laser beam into a line on the facet of the polygon. The line image on the facet should be perpendicular to the axis of rotation. After reflecting from the facet, the line of light from the facet must be restored to a round beam with additional cylindrical optics. This system has been successfully used in laser printers which can write around 300 dots per inch over normal document widths. The cylindrical optics are not conducive to diffraction-limited optics, and for applications requiring more dots per inch, the problems become increasingly difficult. Users are now seeking to achieve 2000 and above dots per inch over large formats. There is also a constant search for increased scanning speeds. For the ultimate polygon scanners, the trend is towards diamond-turned-polygons on-air bearings. These are expensive. They are made on expensive equipment and unfortunately are not made in large numbers. When the cylindrical lens approach is abandoned, the facet must be large enough to cover the entire circular beam. This makes the polygon thicker, heavier, and more expensive to machine, for fewer can be placed on a single mandril.

The polygon scanner facets rotate about a different center than the facet center. This results in a displacement of the mirror as it scans off-axis images. This has to be taken into account in the design of the lens when the cylindrical lenses are used, because the line image on the facet goes out of focus as the polygon rotates. Oddly enough, it actually helps to push the tangential field back to a flat focal plane. If it is left out of the calculations, the lens design will be incorrect. To do this correctly with one of the modern lens design programs, it is best to set the lens up as a multiconfiguration system where the off-axis images are separate configurations.

3.6.3. Holographic Scanners

These scanners use highly efficient transmitting gratings to bend the light. Identical gratings are positioned on a circular disc of glass forming the equivalent to the facets on a polygon. The gratings are all on one plane. The holographic scanner has the advantage that it is much less sensitive to axis-of-rotation wobble. It also has no equivalent to pyramid error. Because of the geometry of its use, the hologon has bow in the scan line. This has to be reduced to the lowest possible value, which is a complex geometrical diffraction grating equation problem that has been solved. The holographic scanner is relatively new compared to the other two scanning systems, but it appears to be ideal for many applications requiring small spot diameters and large formats.

3.6.4. The Design of Lenses for Laser Scanners

As mentioned above, the design of optical scanners is similar to the design of eyepieces or unsymmetrical landscape lenses. The usual design uses a negative lens next to the scanning element. The entrance pupil of the system is the scanning facet. There must be enough space between the scanner and the lens to allow for the moving mirror. The negative lens helps provide some Petzval correction and is a source of positive spherical aberration which introduces hard-to-come-by positive astigmatism. Then there are positive lenses to focus the axial image. The total bending of the chief ray is determined primarily by the distortion required to meet the F-theta condition. The spherical aberration is usually slightly undercorrected, so the focal plane can be focused closer to the lens than the paraxial focus. This helps flatten the field. Most of the scanning lenses used for the 300-dots-per-inch scanning use large F/number lenses. They are often around $F/80$. At this F/number, two and three elements are adequate, provided the scan-angle requirement is not too great. Designers usually want to use as large a scanning angle as possible in order to keep the laser beam diameter small. This keeps the facet size down. The resolution (spot size) requirement determines the F/number. The scan length and field angle are related to the focal length. The result is that large field angles keep the facet size small. Unfortunately, large field angles are troublemakers, particularly when there is an F-theta condition to meet. The distortion curve for this type of lens follows a quadratic curve for third-order distortion which can be matched fairly well to a linear curve, but as the field angle increases, the high-order distortion comes in, and with a simple negative-positive lens system, it tends to be negative. To fit a linear requirement for distortion requires that some source of positive high-order distortion must be found. This requires inserting a surface that deflects the chief ray away from the axis. Stronger negative or thicker positive lenses are required, so that they can be bent into meniscus lenses. To be effective on distortion, the surface should be placed among the positive lenses, which works against attempts to correct the Petzval curvature. It is complicated to understand in detail, but to make a long story short, it can be said that the lens designer should always try to keep the field angle small, and the mechanical designer will attempt to make it large. Good communication and understanding of each other's problems is a great asset.

The above statements become more and more important when the relative aperture of the scanning lens is increased in order to achieve spot sizes.

Figure 29 shows the lens and specifications for a five-element scan lens which is corrected for the F-theta condition. It covers a half-field angle of $30°$ and works at $F/23.8$. The focal length is 100 mm. It forms 3000 independent image points over a scan length of 106.6 mm and follows the F-theta law to within 0.2% over the scan length. The energy concentration in the independent spots across the field is not as good as one would like. However, this is a patented lens and it may not represent the ultimate design. An

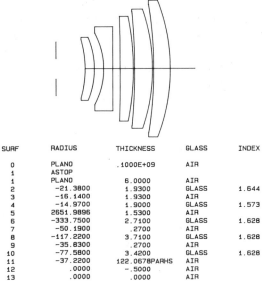

Fig. 29. A wide-angle five-element scan lens with F-theta correction. It images 3000 spots over a scan length of 106.6 mm. (Maeda-Yuko USP 4,269,478.)

SURF	RADIUS	THICKNESS	GLASS	INDEX
0	PLANO	.1000E+09	AIR	
1	ASTOP			
1	PLANO	6.0000	AIR	
2	-21.3800	1.9300	GLASS	1.644
3	-16.1400	1.9300	AIR	
4	-14.9700	1.9000	GLASS	1.573
5	2651.9896	1.5300	AIR	
6	-333.7500	2.7100	GLASS	1.628
7	-50.1900	.2700	AIR	
8	-117.2200	3.7100	GLASS	1.628
9	-35.8300	.2700	AIR	
10	-77.5800	3.4200	GLASS	1.628
11	-37.2200	122.0678PARHS	AIR	
12	.0000	-.5000	AIR	
13	.0000	.0000	AIR	

extra lens with some lens thickness might improve the performance. This lens meets the F-theta condition well. It is probably accomplished by placing the positive lens next to the stop. This positive lens and the following negative lens are probably balancing out the third-order and the fifth-order distortion. Scanning lenses of this type may have some assembly problems resulting in tilted field curves in one meridional plane, but by rotating the lens it is possible to find a flat field. Making the lenses round appears to be wasteful of glass because only a small cross section of the lens is used for the scanning, but several lenses can be salvaged by finding the optimum cross section. Some of these scanning lenses have been sliced in order to keep the size and weight down. This, however, is not recommended for small runs because the tooling costs are high. If the lens is to be made in large numbers, it will pay to do this, but it takes high-quality glass-grinding equipment and expensive mechanical tools in the assembly room.

Figure 30 shows a five-element laser scan lens which meets requirements close to the optical limits of scanning lenses. The lens is corrected to be telecentric and F-theta. The focal length is 48 mm and it operates at $F/3$ at the wavelength of 4416 nm. This lens images 6700 spots on a 16-mm scan length.

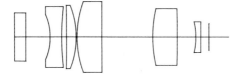

SURF	RADIUS	THICKNESS	GLASS	INDEX
0	PLANO	.1000E+09	AIR	
1	ASTOP			
1	PLANO	16.6000	AIR	
2	PLANO	7.0000	SCHOTBK7	1.526
3	PLANO	14.5770	AIR	
4	-33.4380	6.8905	SCHOTBK7	1.526
5	154.0699	3.0000	AIR	
6	-701.4150	5.6993	SCHOTSF6	1.844
7	-50.304	.5000	AIR	
8	58.9203	14.9983	SCHOTSF6	1.844
9	1035.3423	30.9765	AIR	
10	72.4029	14.9000	SCHOTSF6	1.844
11	-502.7755	11.1577	AIR	
12	-36.6591	3.0000	SCHOTBK7	1.526
13	784.1693	4.9064	AIR	
14	.0000	.0000	AIR	

FIG. 30. A large numerical aperture laser-scanning lens, F-theta, telecentric, $F/3$. This lens images 6667 spots on a 16-mm scan length. (Designed by R. E. Hopkins and Berlyn Brixner.)

The lens wants to have thick elements. When the optimization has no limits on lens thickness, they become larger and larger. Lenses like this are difficult for the designer to understand. As mentioned above, scan lenses are like eyepieces. It is always easiest to correct the Petzval curvature by placing negative lenses at or near the lenses' focal points, for the negative power does not effect the focal length. In wide-angle scan lenses, it is not practical to place a negative lens in the image plane, because the format is usually too large. In the case of a telecentric lens, the first focal point is located at the entrance pupil which is located on the scanning element, so it is not possible to place a negative lens in that position. The lens shown in Fig. 30 fortunately has a small-enough format to make it practical to place a negative lens close to the focal plane. The lens shown may be a bit too close to the image plane: dust particles may disturb the image. It would be necessary to keep the surfaces of this lens clean.

The design of scan lenses for large numbers of spots on a single scan line (some modern requirements are calling for 2000 lines per inch) is relatively new in the lens-design field. As a result, the design limits are not well understood. Up to this point, the following guidelines can be offered.

The telecentric condition, the F-theta, and small spot sizes are conflicting requirements. Careful considerations should be given to trading off these requirements.

There is no point in attempting to design for a large number of scanning spots if one is not prepared to verify that the lens can be built and actually perform to the design. It is a difficult test to determine the source of any errors without adequate interferometric equipment. When the number of spots gets up into the range of 15,000 spots on a scan, the number of shops that can make the lenses goes down rapidly.

High-quality scan lenses are possible because they usually are corrected for single wavelengths. Their performance is rapidly reduced when there is a spectral band width. As with eyepieces, it is difficult to achromatize the off-axis images. It can be done, but it takes many cemented lenses of different glasses.

So far, scanning lenses have been relatively easy to design and manufacture, but as the spot sizes go down and the scan lengths go up, the designs will be much more demanding for the designer and the optical and mechanical shops. However, if developers make the investment in these new lenses, they will have systems that do not wear out, do not need new vacuums or lubrication, but will provide their amazing performance year in and year out for as long as the mechanical and electrical systems continue to provide a good beam and scanning surface.

References

1. S. Rosin, "Eyepieces and Magnifiers," in *Applied Optics and Optical Engineering*, Vol. 3, Chap. 9, Academic Press, New York, 1965
2. R. E. Hopkins, "Eyepieces," in *Military Standarization Handbook: Optical Design, MIL-HDBK* 141, U.S. Defense Supply Agency, Washington, D.C., 1962.
3. J. R. Benford, "Microscope Objectives," in *Applied Optics and Optical Engineering*, Vol. 3, Chap. 4, Academic Press, New York, 1965.
4. A. D. Clark, *Zoom Lenses*, Monographs on Applied Optics, Adam Hilger, 1973.
5. Ellis I. Betensky, "Photographic Lenses," in *Applied Optics and Optical Engineering*, Vol. 8, Academic Press, New York.
6. R. Kingslake, *Lens Design Fundamentals*, Academic Press, New York, 1978.
7. G. Wooters, and E. W. Silvertooth, *JOSA* **55**, No. 4 (1965).
8. J. C. Urbach, T. S. Fisil, and G. Starkweather, "Laser Scanning and Electronic Publishing," *Proc. IEEE* **70**, 597–618 (1982).
9. C. J. Kramer, "High-Speed Read/Write Techniques for Advanced Printing and Data Handling," *Proc. SPIE* **380**.
10. R. E. Hopkins and J. Buzawa, "Optics for Laser Scanning," Opt. Eng. **15** No. 2 (1976).
11. J. Buzawa, "Lens Systems for Laser Scanning," *Laser Focus*, September (1980).
12. G. F. Marshall, ed., *Laser Beam Scanning*, Marcel Dekker, 1985.

4. BASIC OPTICAL INSTRUMENTS

Douglas S. Goodman

IBM T. J. Watson Research Center
Box 218
Yorktown Heights, New York 10598

4.1. Introduction

It is assumed that the reader has an elementary knowledge of geometrical optics and diffraction theory. A detailed knowledge of aberrations and their correction is not assumed, but optical instrument users should at least be aware of their existence and importance. Unless otherwise indicated, statements about instrument performance assume aberration correction.

This survey is limited to laboratory instruments used in and near the visible region of the spectrum. Many types of instruments are omitted; these include astronomical, cinematic, colorimetric, fiber, medical, military, ophthalmic, photogrammetric, and surveying devices. For more information on these, see Kingslake,[1,2,3] Horn,[4] and Glazebrook.[5] We do not discuss the numerous combinations of those instruments that are considered—microscopic spectrometers, spectroscopic ellipsometers, and so on. The camera is excluded, since in the laboratory it usually occurs only in simple and fixed forms as an attachment for microscope or oscilloscope photography. Two useful general references on the camera are Kingslake[6] and Cox.[7] Interferometers are treated elsewhere in this volume. Almost every natural science has its own optical instrument literature, where many devices are treated more thoroughly than in the literature of optics proper.

This chapter is organized principally by the structure of instruments, rather than by application, refractometers being the exception. The sections are not completely independent, but may draw upon those that precede, as well as on other chapters in this volume.

The basic structure and theory of each type of instrument is discussed, as well as performance criteria, limitations, and caveats. There are representative performance figures, intended to give a general idea of capabilities, which are neither maximum performance limits nor levels that can be taken for granted. No attempt has been made to include the most recent developments, which can be found in the scientific and manufacturers' literature. We do not discuss the nonoptical aspects of instruments, such as their detectors, electronics, illumination sources, and mechanical structures.

Simple thin lenses are shown in most of the schematic diagrams for clarity, although they would rarely provide the required aberration correction. Most diagrams are for totally refractive systems, but, in general, there are reflective equivalents.

Because of the long history of optical instruments, their terminology is often confusing, and there is a lack of standardization. The nomenclature is drawn from English, French, and German, as well as from proper names and from trade names. There are often several names for the same thing and several things with the same name. Older terms may reflect erroneous physical theories. Many of the definitions herein are not absolute, but are based on my understanding of common usage. The term "magnification" is particularly troublesome, since a distinction is not usually made between the several types of magnification: transverse, longitudinal, angular, and visual. The term "power" is used in a variety of very different ways.

Some general caveats are appropriate for experimental physicists who use combinations of instruments, or who may "cannibalize" parts. The mating of instruments requires, at the least, proper alignment. Additional components may be required to preserve light flux or field of view, and even with proper mating, there may be significant losses. For instruments in series, transmission factors due to absorption and scattering are multiplied. With respect to geometrical limitations, such as apperture size and acceptance angles, a combination of systems is no better than the worst of its components.

Mixing components and "customizing" should also be done advisedly. Optical elements are specifically designed for the instruments of which they are a part, and they may perform poorly elsewhere. Many components are not individually corrected for aberrations, but have aberrations that balance those of another portion of a system. In addition, the aberrations of a lens depend on the direction it faces and on the magnification at which it is used. These remarks also apply to surplus components of unknown origin.

4.2. The Visual System

Human vision is of interest both for its own sake and because it is through the eye that we obtain most of our knowledge of the physical world. With an appreciation of vision, we know what credence to give our observations, how to use sight most accurately to make judgments, and how to best design visual instruments. The more extraordinary abilities of the visual system should be recognized so that we appreciate the difficulty of matching

THE VISUAL SYSTEM   121

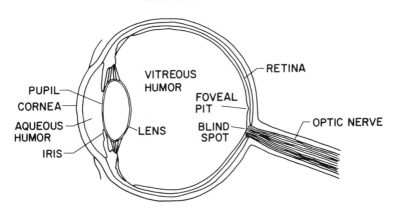

FIG. 1. A cross section of the eye.

them with automata. It is the eye proper that is the optical system, but here we consider the visual system as a whole, since this book is addressed to those who are not vision specialists. There is a vast literature on vision, and the reader is referred to a number of review articles, for instance those by Fry[8,9] and Alpern,[10] and books, for example those by Graham,[11] Pirene,[12] Cornsweet,[13] and Brindley.[14]

There is no single model or small number of models that accounts for all visual phenomena. The limits of vision can sometimes be related to physical structures in the eye, but in other cases these physiological limits seem to be exceeded. We can do little more here than list in a simplified way a number of the properties of the visual system. Omitted entirely are those that have little to do with the typical experimental physics environment, for instance motion perception and depth perception. Individual variations are not considered, and we speak only of the "average" eye operating in "average" conditions, although visual capabilities depend upon illumination levels, color, ambient conditions, etc.

Figure 1 is a cross-sectional diagram of the eye, and Fig. 2 is a schematic showing its cardinal points. The principal planes are near the front surface of the eye, since about two-thirds of the refraction occurs at the cornea–air interface. The nodal points are near the center, since the eye is nearly round. The flexible lens serves principally to adjust focus for viewing objects at different distances. The unit of power in optometric work is the diopter, or inverse meter; a one-diopter lens has a one-meter focal length, and a two-diopter lens has a focal length of half a meter.

The *near point* of the eye is the closest position at which it can focus; the *far point*, the most distant. The standard near point is taken to be at 250 mm, which is said to be the average for a 40-year-old person. The

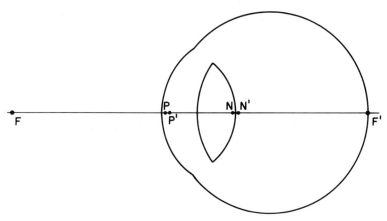

FIG. 2. The schematic eye, showing the approximate locations of its cardinal points. The focal points are F and F'; the principal points, P and P'; the nodal points, N and N'.

standard far point is at infinity. Accommodation is the ability of the eye to focus at different distances, with the standard eye having 4 diopters of accommodation. In nearsightedness, or myopia, either the eye's power is excessive or the eyeball is too long, so that distant objects cannot be focused. If the far point is x meters in front, there is $1/x$ diopters of nearsightedness, which requires for its correction a spectacle lens whose focal length is $-x$ meters. In farsightedness, or hyperopia, the power is insufficient to focus objects at the standard near point. An eye with $1/x$ diopters of farsightedness has a far point x meters behind the eye, which is corrected with a lens of focal length $+x$ meters. A spectacle lens should translate the image axially to the retina without changing the image size. This requires that the lens be placed in the front focal plane of the eye, so that the focal length of the combined system equals that of the eye alone, but the principle planes are shifted.

The pupil of the eye varies in diameter from 2 mm to 8 mm, depending upon the average illumination level, but this change accounts for only a small fraction of the eye's dynamic range. For a 3 mm pupil, the central diameter of the Airy disc subtends 1.5 arc minutes. For greater pupil openings, the potential decrease in the Airy disc diameter is offset by increased aberrations.

The image formed by the eye lies on the retina, where it is detected by receptor cells, the rods and cones. The average center-to-center distance between these cells is 2–2.5 μm, and the angular subtense from the rear nodal point is about 25 arc seconds. Color and high-acuity vision involve the fovea, which contains mainly cones. The high-acuity field of view is about 1°, and the eye constantly sweeps the visual field, directing the fovea to the area of interest. This constant motion also avoids saturation and precludes awareness of the blind spot at the junction of the retina and the optic nerve. Outside the fovea, the predominant receptors are rods, which are more sensitive to light and individually insensitive to color. The cones in the fovea are somewhat smaller than the receptor cells outside, but the greater acuity of the fovea is due principally to its "wiring"; the nerve-to-receptor ratio here is nearly unity, considerably greater than that elsewhere.

A rod can be activated by a single photon, but the activation of a single rod is not perceived. A continuous source of light can be detected if it provides about 5×10^{-10} ergs/second at the wavelength of greatest sensitivity. The energy required for detection is minimized when it is received in a small angle (1 arc minute) and over a short length of time (0.1 second), so that its effects are summed by the retina and the succeeding nerves. The threshold energy for perceiving such a pulse at a wavelength of 0.5 μm is 4×10^{-10} ergs, which corresponds to about 100 photons entering the eye, of which, due to various losses, 5 to 15 are finally absorbed.

The dynamic range of the eye is extraordinary. The ratio of the flux from the sun to that from the weakest detectable star is about 10^{14}. The eye is incapable of judging absolute intensity level, but it is useful as a comparator. According to *Weber's Law*, a fixed percentage change in illumination of level of about 1–2% can be detected over a wide range of illumination levels. Two fields can be most accurately compared if they are juxtaposed. One device that does so is the Lummer-Brodhun head Fig. 3, consisting of two glass components, which are brought into optical contact to sharply separate the fields to be compared. Walsh[15] discusses many such photometric instruments. An instrument that takes advantage of the eye's comparison

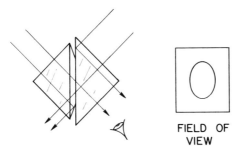

FIG. 3. The Lummer-Brodhun cube for juxtaposing two fields of light. The device consists of a simple prism and a second prism, on one side of which is a spherical surface with a flat portion. The two elements are optically contacted so that part of the surface transmits and the rest provides total internal reflection. Thus the viewer sees light from one direction in the center of the field and from the other direction on the outside of the field.

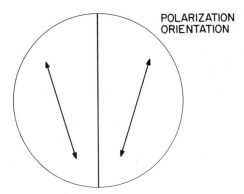

FIG. 4. The half-shade device for finding the azimuth of a plane-polarized beam is constructed of two adjoining polarizers, symmetrically arranged. When the intensities observed on the two sides are identical, the field lies along the bisector.

abilities is the *half-shade* device[16] in Fig. 4, consisting of a pair of side-by-side polarizers, whose directions make equal angles to their common edge. The azimuth of a linearly polarized beam is found by rotating the half-shade unit so that the two fields are of equal brightness. An accuracy of 0.01° is possible, which is far greater than can be obtained by rotating a single polarizer and estimating the position at which the transmitted light is maximum or minimum, since the maximum and minimum orientations are extrema with respect to angle, and because the eye is not good at judging irradiance levels.

Visual acuity[17] is the ability to discriminate detail. Acuity may be specified in terms of object size, retinal size, or in angular or inverse angular units. (At the near point, an arc minute is subtended by about 75 μm or 3 mils, the diameter of a typical human hair. An arc second is subtended by about 1.2 μm or 1/20 mil at the near point, or by a 3 mil hair at 50 feet.) Quoted acuity limits are often those at which experimental subjects make a correct choice in 50% of the tests, so optical instruments should be designed, if possible, with some leeway. These are four categories of acuity: detection, recognition, resolution, and vernier. Acuity test patterns for these categories are shown in Fig. 5.

The *detection* of a bright subresolution object, such as a star, depends only on its flux and is not a matter of acuity. A dark line at least 30 arc minutes in length can be detected against a light background if its width is 0.5 arc seconds. The retinal image of such an object is uniform, except for a drop of about 1%, which corresponds to the intensity discrimination of the eye. Dark circles with diameters of 30 arc seconds and squares with sides of 14 arc seconds can be detected.

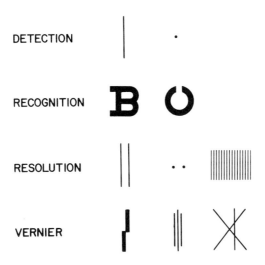

FIG. 5. Sample test patterns for measuring the four types of visual acuity.

Recognition is measured by the familiar Snellen eye chart. A letter can be thought of roughly as consisting of a five-by-five array of regions that must be recognized for its identification. Another test pattern is the Landolt ring, for which the orientation of a break whose width is 1/5 of the diameter must be determined. An acuity of 1 arc minute is typical, so a letter subtending 5 arc minutes can be identified. A score on such a test, of, say, 20/30 means that an object subtending 1 arc minute at 30 feet can be recognized at 20 feet, or that the acuity is 2/3 of the standard value.

Resolution is the ability to perceive the separation of patterns that would be visible if present singly. Two-point resolution capability is about 1 arc minute. A pair of dark parallel lines on a light background can be resolved if separated by 30 arc seconds. Recognizing that a periodic object is nonuniform depends on its angular frequency and contrast. Figure 6 shows the contrast required, as a function of angular frequency, to perceive the periodicity of an object whose brightness varies sinusoidally with position. A unit contrast square wave grating with a 1-arc-minute period can be recognized as nonuniform. This spacing corresponds to that of about 2 cones, which seems to be the fundamental limit.

Vernier acuity, is the ability of the eye to align objects relative to one another. Two straight lines can be aligned to about 5 arc seconds. A related ability is that of the eye to set a line equally between two fixed lines. Whenever possible, measuring instruments should take advantage of the eye's exceptional ability in this regard.

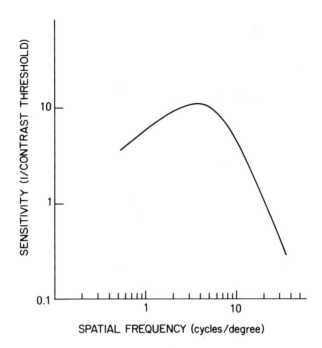

FIG. 6. The visual contrast sensitivity curve, showing the contrast required to detect the modulation of a periodic object as a function of its angular frequency. The object has sinusoidal brightness variation.

Chromatic sensitivity curves for the eye are shown in Fig. 7. The scotopic (dark-adapted) and photopic (light-adapted) curves are normalized to the same value, but they have different meanings. In scotopic vision, only a single pigment is involved, and there is no perception of color, only that of dark and light. Thus its response curve shows the ratios of energy required to obtain similar sensations of brightness, for instance as measured at threshold. In photopic vision, the brightness of, say, red and green stimuli cannot be directly compared, but the relative sensitivities indicate the appearance of mixtures.

The perception of color seems to involve three different pigments. A given color sensation can be caused by any number of spectral distributions. Thus any three different colors can be chosen as primaries, which can be combined to cause any color perception. However, it may be necessary to use subtractive mixing, in which, rather than matching a given color to a sum of the three primaries, two of the primaries are matched to a combination of the given color and the third primary. The eye excels at the comparison of

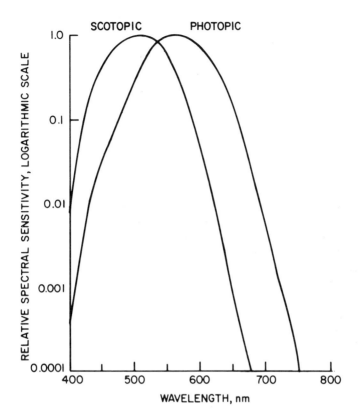

FIG. 7. The normalized spectral sensitivity curves for photopic (light-adapted) and scotopic (dark-adapted) vision.

adjacent uniform color fields, and perhaps 5 million colors can be so discerned.

4.3. Magnifiers and Eyepieces

4.3.1. Magnifiers

A *magnifier* is a single lens that produces for visual observation an erect virtual image of a nearby object, as shown in Fig. 8. The purpose of an *eyepiece* is the same, but its object is an image formed by a preceding optical system. The eyepiece is discussed here as a separate instrument, since its function is the same wherever it is found. Boutry,[18] Rosin,[19] and Kingslake[20] discuss some of the more specialized aspects of eyepiece design and aberrations.

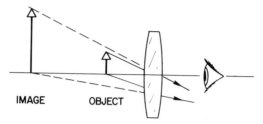

FIG. 8. The simple magnifier forming a virtual erect image of a real object.

As an object is brought nearer to the eye, the size of its image on the retina increases, and it appears larger. The magnification of the eye is limited by the shortest focal length it can have, which determines the closest point at which the object can be seen in focus. By convention, the *visual magnification* or *magnifying power* m_v of a visual instrument is the ratio of the angular subtense of the image to that when the object is located at the standard near point, where it appears as large as possible to the naked eye. The visual magnification is identical to the ratio of retinal image sizes, so it can also be thought of as a ratio of transverse magnifications. The standard eye is most relaxed if the image is at infinity, which is the case if the object is located in the front focal plane of the magnifier. A small object of size h at the near point subtends an angle $h/250$ mm. For a magnifier of focal length f, the subtense of an image at infinity is h/f. The ratio of the two is the visual magnification,

$$m_v = \frac{\frac{h}{f}}{\frac{h}{250 \text{ mm}}} = \frac{250 \text{ mm}}{f}. \qquad (4.1)$$

The subtense of an image at infinity is independent of the axial position of the eye. Another definition of visual magnification also appears in the literature: the image formed by the magnifier is taken to be at the eye's near point, where it has the largest usable subtense. If the eye is at the magnifier, this gives

$$m_v = 1 + \frac{250 \text{ mm}}{f}. \qquad (4.2)$$

For commonly used magnifications, the numeric difference between these two expressions is unimportant, and the former is more often used.

The angle subtended by an image at infinity is

$$\theta = \frac{m_v h}{250 \text{ mm}}. \qquad (4.3)$$

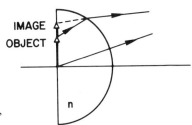

FIG. 9. Hemispherical contact magnifier, for which $m = n$.

For an eye with an acuity of 1 arc minute, the visual magnification required to clearly view object details of size s is

$$m_v \geq \sim \frac{75 \ \mu m}{s} \cong \frac{3 \ mils}{s}. \quad (4.4)$$

Visual magnifications up to about 25× are practical with simple magnifiers. With increasing magnification, the lenses become smaller and more curved, and the eye must be closer.

There are a number of specialized types of magnifiers. Some have a mechanical spacer that fixes the object-to-lens separation, and some have measuring reticles just above the object plane.

An *immersion magnifier* or *contact magnifier* has a flat surface that lies against the object. For a hemisphere (Fig. 9), the image is coplanar with the object, and the transverse magnification equals the refractive index of the lens. The image is aplanatic, that is, free of spherical aberration and coma. It is also free of longitudinal color. The particular hyperhemispherical element shown in Fig. 10, called an *Amici lens*, has a transverse magnification equal to the square of its index. Its image is free of spherical aberration to all orders, as well as coma and astigmatism to third order. (This special element is used often, for instance as the first component in microscope objectives.) Contact magnifiers can be placed against image plane detectors, so that, as seen from object space, the detector appears larger. Thus a given field of view can be covered by a smaller, less noisy detector.[21]

4.3.2. Eyepieces

Unlike a magnifier, which is used for viewing physical objects that scatter light in all directions, an eyepiece is used to view an aerial image, from which light is directed only from the exit pupil of the preceding optics (Fig. 11). The image of the preceding pupil that is formed by the eyepiece is the exit pupil of the entire system. The exit pupil should be a real image, at which the pupil of the eye is located to receive all the light from the optics.

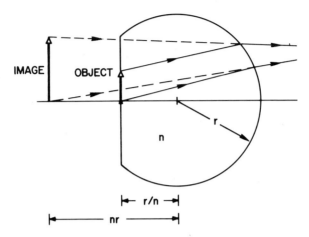

FIG. 10. The aplanatic hyperhemispherical contact magnifier. Its transverse magnification equals the square of its refractive index.

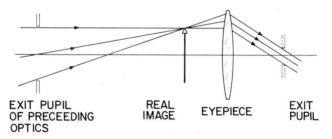

FIG. 11. The eyepiece, essentially a magnifier for viewing the real image, formed by preceding optics. The exit pupil of the entire system is the image formed by the eyepiece of the exit pupil of the preceding optics.

The partial obstruction of some ray bundles, called *vignetting*, unequally darkens the field of view, affecting the outer part most severely. To completely capture the off-axis bundles without an excessively large eye lens, the eyepiece usually incorporates a *field lens* to direct the rays toward the axis. (Fig. 12). In order that its diameter be as small as possible, the field lens should lie in an image plane. In practice it is usually displaced somewhat, so that its surface imperfections and any dirt on its surface are out of focus and not seen. The field lens also allows better aberration correction than would a single lens. Introducing a field lens brings the exit pupil toward the eyepiece. A field lens at the intermediate image does not change the exit pupil size.

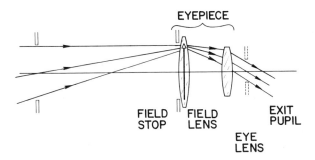

FIG. 12. The field lens, which is placed at or near a real image to reduce vignetting by bending the ray bundles toward the axis.

The exit pupil is sometimes called the *eye circle* or **Ramsden circle**. Its position is often called the *eye point*, and its distance from the rear of the eyepiece is the *eye relief*. A real exit pupil can be located with a sheet of paper or a ground glass if the instrument is viewing a sufficiently bright object. The exit pupil should be either larger or smaller than the pupil of the eye, with the appropriate adaptation, so that the ray bundles are not cut by slight head movements. If stability is a problem, for instance with hand held instruments, a larger pupil is desirable. If the exit pupil is larger than the eye pupil, the eye acts as the system stop, determinining the resolution, so that the image appears as sharp as normal vision permits, aside from aberrations. Because of the optical invariant, the exit pupil size cannot be freely chosen. If m_v is the visual magnification of the eyepiece and d is the distance from the exit pupil of the preceding portion of the instrument to its intermediate image, the ratio of the exit pupil diameter to that of the preceding instrument is $250 \text{ mm}/d m_v$.

In the object plane of an eyepiece, there is usually a *field stop*, which sharply delimits the field of view. This sometimes serves as a cosmetic device to prevent seeing regions in the field where aberrations are uncorrected and where there would be vignetting. The *angular field of view* of an eyepiece is the angle subtended by the image of the field stop.

Reticles, also called *graticules*, are located in the plane of the field stop, coplanar with the real image. There is a variety of reticles for alignment, linear and angular measurement, etc., and there are vendors who make special reticles. Reticles should be calibrated for measurements in the instrument in which they are to be used. If the reticle is to be changed frequently, an eyepiece with an external reticle is preferred. In such eyepieces, the reticle image benefits from the aberration correction of the entire eyepiece. Without a reticle, the field stop can be used as a rough ruler, which can be calibrated by viewing an object of known dimensions. There are eyepieces with illuminated reticles, used for viewing dark objects.

An eyepiece with a reticle should be focused with respect to both the reticle and the image to be viewed. First, the reticle is brought into focus by adjusting the portion of the eyepiece that follows it. Next, the image is focused on the reticle by axially translating the image or the entire reticle. Coplanarity of the image and reticle can be checked by moving the head side to side to check for parallax. Binocular instruments have separate focus adjustments for one or both of the eyepieces to compensate for differences in the power of user's eyes.

The two classical eyepiece designs are the *Ramsden* and the *Huygenian*. The Ramsden is similar to that in Fig. 12, but with the field lens further to the rear. The lenses are plano-convex, with the plano surfaces facing outward. In the Huygenian design, the image is between the two lenses. Because of the location of its object, the Huygenian is often erroneously referred to as a "negative eyepiece," although its power is positive. Numerous eyepiece designs are discussed in the references above and in those in the microscope section. The eyepiece prescribed by the manufacturer should usually be used, since it may be a *compensating eyepiece*, with aberrations cancelling those of the rest of the instrument.

There are several other types of eyepieces. Zoom eyepieces have variable magnification. There are interferometric eyepieces and eyepieces with moveable pointers. The *filar micrometer eyepiece* has a fixed reticle and a moving cross hair, translated by a micrometer.

4.4. Afocal Systems

Afocal systems,* whose power is zero, have a number of unique and useful properties that are not widely known. Afocal systems are also referred to as telescopes, since visual telescopes are afocal. However, the term "telescope" is used in other ways, referring, for example to an objective alone.

Afocal and focal systems differ in kind, not merely in degree; afocal systems are not merely focal systems with an infinite focal length. Afocal systems have no focal points, no principal planes, and no nodal points, that is, none of the cardinal points of a focal system (Fig. 19, Chapter 2). The equations for focal systems [Eqs. (2.36)–(2.40)] relating object and image locations and magnification are inapplicable to afocal systems, since they use distances measured from the cardinal points. Focal length is the reciprocal of power, and afocal systems have zero power, but it is preferable

---

* Reference added in proof.

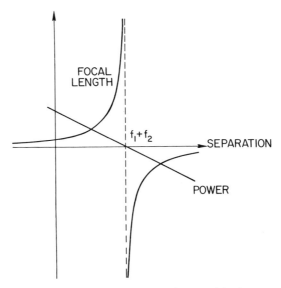

FIG. 13. The power and focal length of a system of two positive lenses as a function of their separation.

to consider the focal length as undefined, rather than as infinite. Figure 13 shows how the focal length and power of a system of two positive lenses vary with their separation. When the power vanishes, the focal length changes from large and positive to large and negative.

The general properties of afocal systems are shown in Fig. 14. Incoming rays parallel to the lens axis are not focused but leave parallel to the axis, in general at a different height. Beam expanders employ this property. The *transverse magnification m* is the ratio of outgoing to incoming heights,

$$m = \frac{y'}{y}, \qquad (4.5)$$

which is the same for all rays parallel to the axis. A parallel bundle of incoming rays at an arbitrary angle leaves as a parallel bundle, but in general at a different angle. It is this property of afocal systems that is used in viewing distant objects. For meridional rays, those in a plane containing the lens axis, the ratio of the tangent of the outgoing angle u' to that of the incoming angle u is the *angular magnification*,

$$m_{\text{ang}} = \frac{\tan u'}{\tan u}, \qquad (4.6)$$

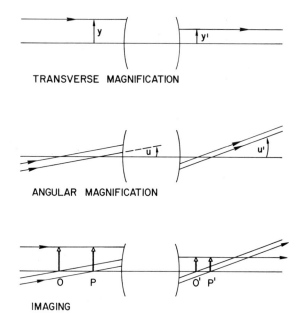

FIG. 14. The general properties of afocal systems, shown here for the case of transverse magnification $m = 1/2$. Rays entering parallel to the axis leave parallel, in general at a different height. Parallel bundles leave parallel, in general at a different angle, as determined by the angular magnification. Images are formed of objects at finite distances.

which is nominally constant for a given afocal system. The view through a telescope is distorted if this ratio varies with field angle. The transverse, angular, and longitudinal magnifications of an afocal system are related by the optical invariant in the same way as for focal systems. The product of the paraxial transverse and angular magnifications is

$$mm_{\text{ang}} = \frac{n}{n'}, \qquad (4.7)$$

where n and n' are the refractive indices of the object and image spaces. The longitudinal magnification, also a constant, equals the square of the transverse magnification,

$$m_{\text{long}} = m^2. \qquad (4.8)$$

Afocal instruments can form real images of real objects at finite distances. The fundamental difference from focal systems is that the magnification does not depend on the object position. To find the image locations for all object locations for a given afocal system, an arbitrary object position is chosen, say O in Fig. 14, and its image O' is located by algebraic calculation or by ray tracing. Using these points as references, the location of the image P' of an axial point P is found by using the longitudinal magnification,

$$O'P' = OPm_{\text{long}} = OPm^2, \qquad (4.9)$$

where OP and $O'P'$ are directed distances. The entrance and exit pupils are a convenient pair of conjugate points. For a two-lens afocal system, a useful pair of conjugate reference points is the front focal point of the first lens and the rear focal point of the second.

4.4.1. Telescopes

Afocal systems are used to view faraway objects; hence the name telescope. For an object so distant that it can be considered to be at infinity, light from the various points of the object arrives in nearly parallel bundles, whose angular separation determines the object's perceived size. To increase its apparent size, the angles between the bundles are increased by the telescope's angular magnification, whose magnitude is greater than unity. The action of the telescope can be thought of in another way. Images of distant objects are smaller than the objects by the lateral magnification m, but are closer by the longitudinal magnification m^2. Thus the angular subtense of the image is greater than that of the object by a factor of $1/m = m_{\text{ang}}$.

The simplest telescope is the *Keplerian* (Fig. 15), comprised of two positive lenses whose separation equals the sum of their focal lengths. (More precisely, the separation is measured between their interior principal planes.) The objective, whose focal length is f_{obj}, forms a real image of a distant object that is reimaged, inverted, at infinity by an eyepiece of shorter focal length f_{eye}. The magnifications are

$$m_{\text{ang}} = \frac{1}{m} = -\frac{f_{\text{obj}}}{f_{\text{eye}}}. \qquad (4.10)$$

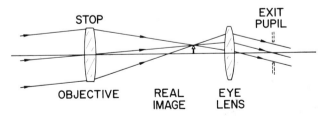

FIG. 15. The Keplerian telescope, consisting of two positive lenses separated by the sum of their focal lengths. The image is inverted.

The objective lens is the aperture stop and entrance pupil, and the exit pupil is the real image of the stop formed by the eyepiece. If d is the objective diameter, the exit pupil diameter is

$$d|m| = d\frac{f_{eye}}{f_{obj}} = \frac{d}{|m_{ang}|}. \tag{4.11}$$

The distance from the rear principal plane of the eye lens to the exit pupil is

$$f_{eye}\left(1 - \frac{f_{eye}}{f_{obj}}\right) = f_{eye}(1 + |m|). \tag{4.12}$$

As discussed above, the simple Keplerian telescope suffers from vignetting, which can be eliminated by adding a field lens.

Telescopes with positive angular magnifications give erect images and are often called *terrestrial telescopes* (in the heavens there is no "up"). The simplest is the *Galilean telescope* (Fig. 16), which has a negative eye lens, whose distance from the objective again equals the sum of the focal lengths. Equation (4.10) also holds for this configuration. The viewer's eye pupil is the stop, and the entrance pupil may be far to the rear of the objective,

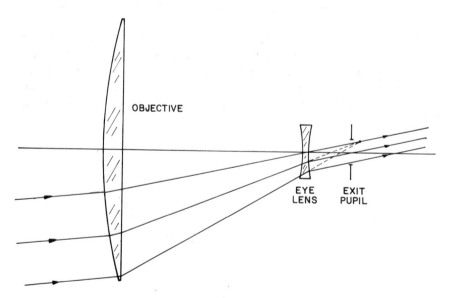

FIG. 16. The Galilean telescope, which has a positive objective, a negative eye lens, and an erect image. This figure shows the stop behind the negative lens, which is the case when the telescope is used visually, so that the eye acts as the system stop.

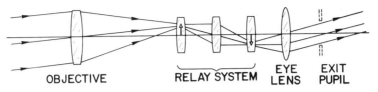

FIG. 17. A Keplerian telescope with a relay arrangement for image erection, consisting of two field lenses and a lens that inverts the real image formed by the objective.

giving considerable vignetting. This design is compact, the total length being less than the focal length of the objective, so it is used in opera glasses.

A terrestrial telescope which does not suffer from vignetting (Fig. 17) is a Keplerian telescope with an inverting *relay* unit. An erect image is obtained at the expense of greater length. Such designs are used, for example, in rifle scopes, where there is the additional requirement of long eye relief.

An image can also be made erect with mirrors or reflecting prisms, as described by Hopkins,[22] Swift,[23] and Kingslake.[3] An even number of reflections is required for correct image parity. Traditional binoculars employ a Porro prism system (Fig. 18). (This effectively increases the interocular separation, so that parallax depth cues are helpful at greater distances.) The more compact "straight through" binoculars may use the prism shown in Fig. 19. Binoculars are specified by a pair of numbers, for instance 7×35, which are the angular magnification, here 7, and the objective (and stop) diameter in millimeters, here 35 mm. The exit pupil diameter is the quotient of the two, here 5 mm, since the transverse magnification is 1/7. Binoculars for night use have large objectives and large exit pupils, which, however, provide no benefit in daylight, when the eye pupil is smaller.

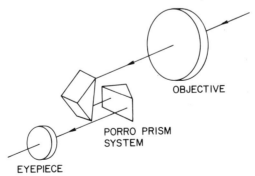

FIG. 18. Half of the traditional binocular arrangement, which is like a pair of Keplerian telescopes, whose images are erected by Porro prisms. (These prisms are usually cemented, but are separated here for clarity.)

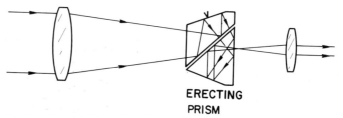

ERECTING PRISM

FIG. 19. A type of straight-through prism erecting arrangement. The "V" in the diagram is a roof, which is required to give an even number of reflections, and hence correct parity.

4.4.2. Properties of Afocal Systems

Most of the finite conjugate properties and applications of afocal systems have to do with their constant magnifications. Tilted objects are imaged anamorphically (Fig. 20), that is, with different magnifications in the two principle meridians, but without the keystone distortion characteristic of focal systems. For magnifications of +1 and −1, the imaging is not anamorphic.

Afocal systems with $m = \pm 1$ have unit longitudinal magnification, so the object-to-image distance is fixed, and a longitudinal shift of the optics does not move the image. Afocal systems with $m = 1$ have geometrically identical object and image spaces, and the image is unaffected by translating the optics in any direction. (Tilting the optics does tilt the image). Thus a unit magnification afocal system can be used as a scanner between a master and a copy, which is duplicated as if by contact printing. This principle is important in photolithography, since such a projector with high resolution

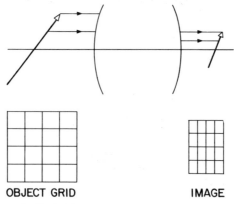

OBJECT GRID   **IMAGE**

FIG. 20. The image of a tilted object formed by an afocal system has no keystone distortion, but the magnification varies with orientation. Squares whose sides are parallel to the two principal directions are imaged as rectangles.

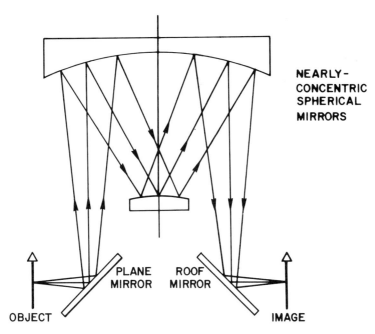

FIG. 21. Schematic of the Perkin–Elmer Micralign ®, an all reflective, unit power, doubly telecentric afocal system used in photolithography. The object (a mask) is imaged onto a wafer, and the two move together on the same carriage to maintain registration. One of the mirrors is a roof to give an even number of total reflections, so that the object and image spaces have the same parity.

over a small "instantaneous field of view" can copy large objects (Dyson,[24] Wynne,[25] Offner,[26] and Lin[27]). This application also takes advantage of the unique feature of afocal systems that they can be telecentric in both object and image spaces. One such printer, the Perkin–Elmer Micralign® Model 200, is shown schematically in Fig. 21.

In another application of the afocal system (Fig. 22), two scanning mirrors at conjugate positions are effectively superimposed, so that a beam can be deflected in any direction from a fixed vertex. The relative sensitivities of the mirrors depend on the magnification of the system. A field lens can be added to reduce the diameters of the outer lenses.

As with the Keplerian telescope field lens, the addition of a lens at the common focal point of a two-lens afocal system leaves the system afocal and the magnifications unchanged, but it does change the positions of conjugate pairs. Any object-to-image distance can be obtained, as Fig. 23 shows for an $m = -1$ Keplerian system. An application of this device for alignment is given by Burch and Williams.[28]

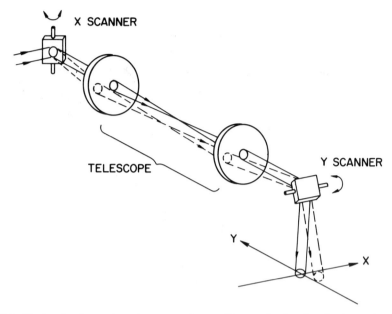

FIG. 22. A pair of scanning mirrors superimposed by an afocal system to give uncoupled x and y deflections about a fixed vertex at the final mirror.

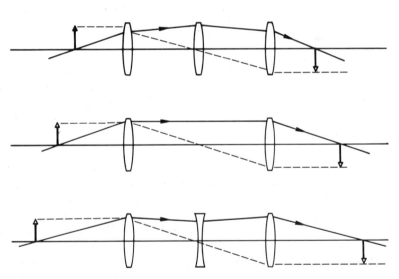

FIG. 23. A set of afocal systems with identical magnification, but different object-to-image distance, which depends on the power of the central lens. The central thin lens does not affect the dashed ray shown in the central diagram, so afocality is maintained.

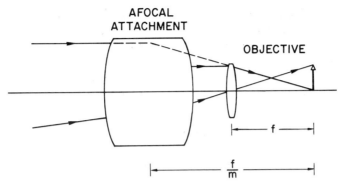

FIG. 24. An afocal system used to increase the focal length of a lens without moving its rear focal plane. Such devices are used as attachments for photographic lenses.

Placing an afocal system in front of a focal system (Fig. 24) results in a focal system with a different focal length, but the location of the rear focal plane is unchanged. If f is the original focal length and m is the transverse magnification of the afocal system, the focal length of the combination is f/m, regardless of the axial position of the afocal system. This principle is used in photographic attachments, and a visual telescope can be thought of in this way as increasing the focal length of the eye.

We have discussed afocal systems only in terms of ray theory. From the viewpoint of wave theory, their fundamental property is that incident plane waves leave as plane waves, aside from the effects of finite apertures and of aberrations. For $m = \pm 1$ the angles between the various plane wave fronts are unchanged on passing through the system. So for an object illuminated with a plane wave, the plane-wave spectrum of the image is nominally identical to that of the field just beyond the object, an important property in many coherent imaging processing systems,[29] as shown in Fig. 25. When

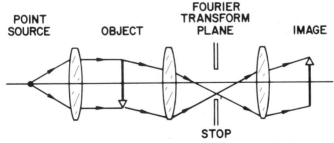

FIG. 25. An afocal coherent processing system. Masks in the Fourier-transform plane perform spatial filtering. The rays shown are those associated with a single-plane wave component of the object field.

used symmetrically, with the object in the front focal plane of the first lens and the image in the rear focal plane of the second, a two-lens afocal system also provides a Fourier transform plane with no phase curvature at the rear focal plane of the first lens. Because of vignetting and aberrations, other arrangements are usually preferred if phase curvature is unimportant, for example, if only the irradiance of the image is of consequence.[29]

4.5. The Autocollimator

The autocollimator, one of the most important instruments in metrology, is used both for measuring small differences in angles and as a nulling device for setting angles. Hume[30] discusses the basic principles and many applications (for which see also Mauro,[31] Young,[32] Habell and Cox,[33] and other books on metrology and optical testing). Two more recent discussions are those of Torberg et al.[34] and Bottomley,[35] who treat photoelectric devices, which are omitted here.

The basic principle of the autocollimator is shown in Fig. 26. A reticle centered in the focal plane of a telescope objective is illuminated by means of a beam splitter and an offset lamp, and is viewed with an eyepiece. (This unit is called a *Gauss eyepiece.*) The light transmitted by the reticle passes through the objective, is reflected by a plane specular surface, and an image of the reticle is formed in the reticle plane. If the reflector surface is perpendicular to the lens axis, the reticle is imaged upon itself, inverted. If the reflector is tilted by a small angle θ, then regardless of the distance of the reflector, the reticle image is displaced by

$$y = 2\theta f, \qquad (4.13)$$

where f is the focal length of the objective. From Eq. (4.6), the two images

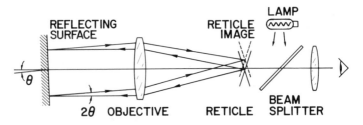

FIG. 26. The principle of the autocollimator. The reticle is imaged onto itself, with a shift that depends on the angle that the reflecting surface makes with the instrument axis. The distance to the reflecting surface does not matter, since the reticle is at the principal focal plane of the objective.

of the reticle seen through the eyepiece have an angular displacement of

$$\alpha = 2\theta m_{ang}, \qquad (4.14)$$

where m_{ang} is the angular magnification of the telescope.

Unlike the simplified device of Fig. 26, most instruments, such as that in Fig. 27, have a fixed illuminated reticle that is imaged upon a second reticle, which can be translated by a micrometer calibrated in angular units. The reticle patterns are designed to make best use of the vernier acuity of the eye. A *one-axis* autocollimator has a single moveable reticle, and a *two-axis* instrument has a pair at right angles. To permit measurement in any orientation, the reticle unit may be capable of rotation. In a dark-field autocollimator, the reticle is seen as light against a dark background.

The principal performance specifications of autocollimators are precision, accuracy, angular range, and maximum working distance (100 to 150 feet is typical). Angular range decreases with precision. The measurement capabilities claimed by manufacturers are typically from 1 arc second to 0.1 arc second, but just what is claimed is often unclear, since the terms used include "resolution," "sensitivity/repeatability," "setting accuracy," "minimum increment," and "sensitivity." This confusion may arise because autocollimators are used both as comparators, for which precision or repeatability should be used to specify performance, and as small-angle measurement devices, where accuracy is appropriate. External conditions and operator skill affect performance.

Some typical applications are shown in Fig. 28. In (a), the flatness of a surface plate is measured by the changes in the angle of a mounted mirror traversing it. In (b), the angle of a corner reflector is checked; a pair of reticle images is produced if this angle is not 90°. Arrangement (c) shows how separated surfaces can be set perpendicular by comparing the reticle images with and without a precision pentaprism, which deviates the beam by 90° regardless of its orientation.

Telescopes in optical instruments often do double duty as autocollimators. In a prism spectroscope, for example, the telescope may be so configured to align the prism and to measure its vertex angle.

4.6. Projectors

The term *projector* is used here in a general way to include all optical systems that both illuminate an object and image it (Fig. 29). This use of "projector" is not standard, but it seems to be the best term to encompass this class of instruments. We include systems that image transparent objects, as well as those systems whose objects are opaque and illuminated from

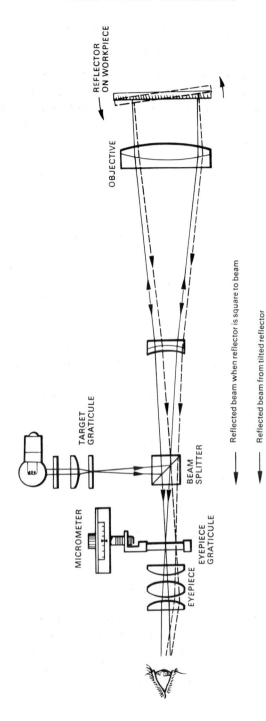

FIG. 27. An autocollimator (Courtesy of Rank Precision Industries, Des Plaines, Ill.).

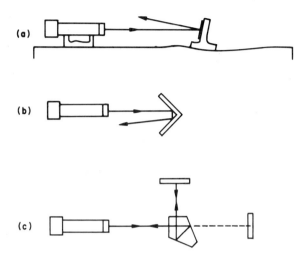

FIG. 28. Some autocollimator applications: (a) checking surface flatness by measuring the angle of a carriage traversing the surface, (b) checking the squareness of a corner reflector, (c) setting separate planes perpendicular with the help of a penta prism, which deviates light by 90°, regardless of its azimuthal orientation.

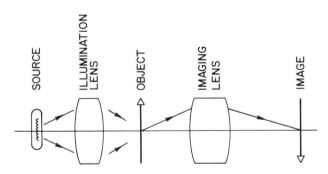

FIG. 29. General schematic of an optical projector, consisting of a source, illumination optics, object, imaging optics, and image.

the same side as the imaging optics. The diagrams of opaque-object projectors can be unfolded about the object plane, and they can be thought of in the same way as transmitting systems. The instruments discussed here include most microscopes, schlieren instruments, photographic and photolithographic copiers, comparators, many coherent processing devices, and many spectrometers, which is to say most imaging systems used in the laboratory. These instruments have evolved independently and are usually treated separately, but they can be thought of in the same general way. Because of the importance of projectors and the shortage of general references, this section is somewhat different than the others. Since many treatments of imaging theory are almost exclusively mathematical, an attempt has been made here to supply some of the needed heuristics.

The properties of a projector are determined by the combined effects of its illumination and imaging systems. The illumination system does not simply supply light, but it affects the character of the image. Geometric optics gives the image location and magnification, as well as the image irradiance with no object present. At low resolution, some illumination effects, such as shadows and specular reflections, can also be treated geometrically. However, a more precise description of the image requires diffraction theory and the theory of partial coherence, which are combined in the theory of partially coherent imaging, as discussed by Hopkins,[36] Hopkins,[37] Thompson,[38] Born and Wolf,[39] Marathay,[40] Singh and De,[41] Goodman,[42] and Goodman.[43] This theory includes the limiting cases of coherent and noncoherent imaging. Historically, the theory of partially coherent imaging arose first in connection with the microscope, since illumination in microscopy is under the control of the user and is seen to play a major role. (This is in contrast to astronomy, where the objects are self-luminous.) Thus, the so-called theory of image formation in microscopes,[44] due to Abbe, is, in fact, general.

The customary treatment of image formation uses scalar wave theory, which suffices for all but the highest numerical apertures. The treatment is usually quasi-monochromatic, which means that the spectral bandwidth is sufficiently wide so that there can be spatial noncoherence, and sufficiently narrow so that wavelength-dependent effects in the optics or in diffraction angles are negligible. The illumination is described by the complex mutual intensity function (or by the coherence function and the intensity) of the field incident on the object plane. The optical path lengths from each object point to its image are nearly equal. Thus, the full coherence function is not required in considerations of imaging; spatial coherence is important, and temporal coherence per se is not, although the image obviously does depend on the chromatic content of the illumination. (The situation is roughly opposite that of many forms of interferometry, where temporal coherence is of primary importance.) In the simplest case permitting analytic treatment, the object is taken to be thin, so it can be described by a complex amplitude transmittance function, which gives the change in magnitude and phase produced on light passing through it. The imaging system is represented by its complex point response function, the amplitude of a point image. The mutual intensity of the illumination is changed by passing through the object, and it is further modified by the imaging system. From the image plane mutual intensity, the image irradiance is found, which is a bilinear function of the object amplitude transmittance.

For most purposes, only the image irradiance is of interest, and it can be determined without making explicit use of coherence theory.[36] Each point of a self-luminous source makes a contribution to the image that can be found by using the principles of coherent image formation. These contributions do not interfere, and so the net image irradiance is just the sum of the irradiances due to each source point. (Spatially coherent sources are not used, since the light from them would interfere in the object plane, giving nonuniform illumination.)

It is usually desired that projector imaging be stationary (or shift-invariant) with respect to irradiance. In this case, an in-plane translation of the object results in a like translation of the image irradiance, which is otherwise unchanged. This requires that the point response function of the imaging system be unchanged with position, except, possibly, for an overall phase term depending on the position of the point in the object plane and a phase term that depends on the image coordinates. Such an imaging system is stationary when the object is illuminated noncoherently. Further requirements for a stationary projector are that the illumination must be uniform on the absolute dimensions of the two. Alternately, coherence effects depend on a difference in object plane coordinates, except for an overall phase term cancelling that of the imaging system. A necessary and sufficient condition for stationarity is that the image irradiance of a two-pinhole object be independent of its position for all pinhole separations. If a stationary projector is also axially symmetric, the form of an image is independent of the object's azimuthal orientation.

The equations for the image irradiance of a stationary projector are as follows. Here the magnification is normalized to unity, $O(\mathbf{x})$ is the complex object amplitude transmittance, and $I(\mathbf{x})$ is the image irradiance. The quantity $\mathbf{x}$ represents coordinate pairs in the object and image planes. The image irradiance and object transmittance are related by the convolution-like integral

$$I(\mathbf{x}) = \int d\mathbf{x}_1 \, d\mathbf{x}_2 \, C(\mathbf{x}-\mathbf{x}_1; \mathbf{x}-\mathbf{x}_2)O(\mathbf{x}_1)O^*(\mathbf{x}_2). \qquad (4.15)$$

The complete projector, both its imaging and illumination attributes, is accounted for by the function $C(\mathbf{x}_1; \mathbf{x}_2)$, which is given by

$$C(\mathbf{x}_1; \mathbf{x}_2) = H(\mathbf{x}_1)H^*(\mathbf{x}_2)J(\mathbf{x}_2 - \mathbf{x}_1), \tag{4.16}$$

where $H(\mathbf{x})$ is the amplitude response function of the imaging system, and $J(\mathbf{x})$ is the mutual intensity of the illumination. Here the phase terms associated with the imaging and illumination functions are not shown, since these terms cancel in the case of stationary imaging.

Imaging can also be treated in the spatial frequency domain. Here the imaging system is described by its pupil function $\tilde{H}(\mathbf{f})$, the Fourier transform of $H(\mathbf{x})$. The illumination system is described by the transform of $J(\mathbf{x})$, the *effective source* $\tilde{J}(\mathbf{f})$, whose physical significance is discussed below. The projector is represented by the function $\tilde{C}(\mathbf{f}_1; \mathbf{f}_2)$, which is like the double Fourier transform of $C(\mathbf{x}_1; \mathbf{x}_2)$, except for the reversal of the sign of $\mathbf{f}_2$:

$$\tilde{C}(\mathbf{f}_1; \mathbf{f}_2) = \int d\mathbf{x}_1 \, d\mathbf{x}_2 \, C(\mathbf{x}_1; \mathbf{x}_2) \, e^{2\pi i (\mathbf{x}_1 \cdot \mathbf{f}_1 - \mathbf{x}_2 \cdot \mathbf{f}_2)}$$

$$= \int d\mathbf{f} \, \tilde{J}(\mathbf{f}) \tilde{H}(\mathbf{f} + \mathbf{f}_1) \tilde{H}^*(\mathbf{f} + \mathbf{f}_2). \tag{4.17}$$

The Fourier transform of the image irradiance $\tilde{I}(\mathbf{f})$ is related to that of the object amplitude transmittance $\tilde{O}(\mathbf{f})$ by

$$\tilde{I}(\mathbf{f}) = \int d\mathbf{f}' \, \tilde{C}(\mathbf{f} + \mathbf{f}'; \mathbf{f}') \tilde{O}(\mathbf{f} + \mathbf{f}') \tilde{O}^*(\mathbf{f}'). \tag{4.18}$$

An advantage of working in the frequency domain, is that the functions $\tilde{H}(\mathbf{f})$, $\tilde{J}(\mathbf{f})$, and $\tilde{C}(\mathbf{f}_1; \mathbf{f}_2)$ are simpler than $H(\mathbf{x})$, $J(\mathbf{x})$, and $C(\mathbf{x}_1; \mathbf{x}_2)$.

In the frequency space viewpoint, the object is thought of as a sum of gratings of various frequencies, phases, and orientations. Each frequency component $\mathbf{f}$ contributes to the image a uniform image irradiance proportional to $\tilde{C}(\mathbf{f}; \mathbf{f})$, and the total image power is $\int d\mathbf{x} I(\mathbf{x}) = \tilde{I}(0) = \int d\mathbf{f} \, \tilde{C}(\mathbf{f}; \mathbf{f}) |\tilde{O}(\mathbf{f})|^2$. Each pair of object frequencies $(\mathbf{f}_1, \mathbf{f}_2)$ gives a sinusoidal image irradiance contribution of frequency $\mathbf{f}_1 - \mathbf{f}_2$. The modulation and phase of this contribution depends on $\tilde{C}(\mathbf{f}_1; \mathbf{f}_2)$. There is a characteristic type of $\tilde{C}$ for each type of system—dark-field, phase contrast, etc. The expression for $\tilde{C}$ in Eq. (4.17) involving the effective source and pupil functions has a simple physical interpretation. For objects that vary in one direction, the imaging equation can be represented pictorially in a way that reveals many of its implications. These matters are discussed by Goodman.[43]

In the limit of noncoherent illumination, $J(\mathbf{x}) \propto \delta(\mathbf{x})$, where $\delta(\mathbf{x})$ is the Dirac delta. Thus $\tilde{J}(\mathbf{f})$ becomes a constant, and $\tilde{C}(\mathbf{f}_1; \mathbf{f}_2) = Q(\mathbf{f}_1 - \mathbf{f}_2) \delta(\mathbf{f}_1 - \mathbf{f}_2)$, where Q is proportional to the optical transfer function.

In the limit of coherent illumination with a centered point source, $J(\mathbf{x}) =$ constant, and $\tilde{J}(\mathbf{f}) \propto \delta(\mathbf{f})$.

To understand why illumination coherence affects image irradiance, consider an object that is opaque except for two subresolution pinholes (Rayleigh,[45] Hopkins and Barham,[46] Grimes and Thompson[47]). Each pinhole gives rise in the image plane to spots of finite size that overlap and interfere. Therefore, the net irradiance is the sum of contributions due to the pinholes taken individually and an interference term, which depends on the separation of the points and the coherence of the light incident on them. Roughly speaking, the interference effects are greatest if the light incident on the pinholes originates at a single source point, and they decrease as the source size increases. In the pupil of the imaging system there is a complementary field, a fringe pattern whose modulation depends on the illumination coherence. (To treat this object analytically in the above equations, use the object function $O(\mathbf{x}) = a_1 \delta(\mathbf{x} - \mathbf{x}_1) + a_2 \delta(\mathbf{x} - \mathbf{x}_2)$.)

This viewpoint can be broadened to an extended object by thinking of it as sampled by a fine array of pinholes. The net image irradiance consists of two parts. The first is the sum of one-point image irradiances, which depends only on the imaging system, the illumination irradiance, and the object irradiance transmittance $|O(\mathbf{x})|^2$. This portion of the image is identical to that which would be obtained with noncoherent illumination of the same irradiance. The second part of the image is the interference contribution, the sum of cross terms between all pairs of points. It depends on the imaging system, the phase shifts introduced by the object, the illumination coherence, and the illumination irradiance.

Coherence effects in image formation depend on the size of the illumination coherence function relative to that of the imaging spread function, not on the absolute dimensions of the two. Alternately, coherence effects depend on the ratio of the size of the effective source in the aperture of the imaging lens to that of the aperture. If the transverse illumination coherence distance is small compared with the resolution distance and the effective source is large compared to the entrance pupil, then the image formation is nearly noncoherent. In the converse case of large coherence distance and small effective source, the imaging is nearly coherent. In images of two pinholes, the magnitude of the cross term is proportional to the magnitude of the coherence function evaluated at the distance between the holes. (An example of effectively noncoherent illumination is outdoor photography, where the resolvable spot of the camera is much larger than the coherence distance of sunlight, ~ 50 μm.) In the case that is most commonly treated, the source is a uniform circular disc centered on the projector axis, and the imaging lens has a circular pupil. In this case, the illumination mutual intensity function has the form of the Airy pattern $J_1(ar)/r$. For such symmetric

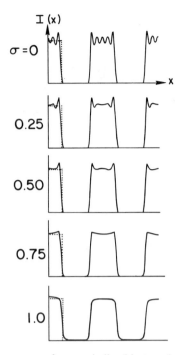

Fig. 30. Image irradiance curves for a periodic object, such as a Ronchi ruling, whose amplitude transmittance is a square wave, a portion of which is shown as a dotted line. In this series, the spatial coherence, which depends upon the parameter σ, is decreased from top to bottom. This shows how images of binary amplitude objects typically depend on the illumination coherence. With lower coherence, the image is more rounded, and the slopes are not as great.

projectors, the ratio σ of the diameter of the effective source to the diameter of the entrance pupil equals the ratio of the diameter of the first zero of the Airy pattern of the imaging lens to the first zero of the Airy pattern of the coherence function. In the coherent limit, $\sigma = 0$; in the noncoherent limit, $\sigma = \infty$.

The manner in which some images vary with illumination coherence is shown in Figs. 30 and 31, which give calculated irradiance profiles for an aberration-free projector with a circular pupil and a circular centered uniform source.[44] The image in Fig. 30 is that of a periodic "amplitude object," one that affects the magnitude of the field passing through it, but not its phase. This object is periodic with alternating dark and clear regions of equal width. Its fundamental spatial frequency is $0.1 \, \text{NA}/\lambda$; so for a centered point source, the $\sigma = 0$ case, diffraction orders up to ± 10 contribute to the image. As the source size is increased, the image irradiance becomes

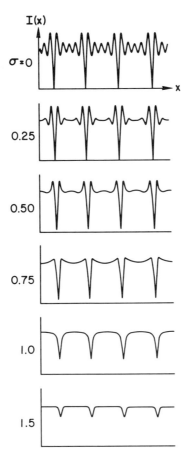

FIG. 31. Image irradiance curves for a periodic-phase bar object with 180° discontinuities. As the spatial coherence of the illumination decreases, the phase discontinuities in the object affect the image less.

smoother and more rounded, and its maximum slope decreases. For $\sigma \geq 1$, the image is very similar to a noncoherent image. Figure 31 shows images of a phase grating of the same period, with alternating strips of equal width that introduce a phase different of 180° and no difference in magnitude. Because this object has no zeroth diffraction order, its image irradiance has twice the frequency of the amplitude object, so that there is a "frequency doubling." (This is so only for the special case of 180° phase shifts and strips of equal width.) The effects of decreased coherence are qualitatively the same as for the pure amplitude object, but as σ increases, the image of an object with discrete phase steps approaches more slowly than does that of an amplitude object its noncoherent limiting form, a field of uniform irradiance. (Most microscopes have an iris that permits the adjustment of illumination coherence, and such phenomena can be readily observed.) A number of papers show how images vary with illumination coherence. For instance, Grimes[48] treats the tri-bar object commonly used in test patterns; Watrasiewicz[49] and Kintner and Sillitto[50] discuss the images of edges.

Illumination must be tailored for each task, and no degree of coherence is best for all purposes. Figures 30 and 31 show some of the considerations. For example, if it is desirable that the image irradiance should be insensitive to object phase effects, such as those due to thickness variations, then the illumination should be as noncoherent as possible. On the other hand, illumination of greater coherence gives irradiance profiles with steeper slopes, which are beneficial for dimensional control in high-contrast lithography.

Image irradiance shift invariance can be obtained with three forms of illumination: Koehler, critical, and diffuse. The most common is *Koehler illumination*, Fig. 32, in which the source is imaged into the entrance pupil of the imaging system, and the object is located in the condenser exit pupil, so the marginal rays of the illumination system are the chief rays of the imaging system and vice versa. In Koehler illumination, each point on the source illuminates the object plane uniformly, and each point on the object plane is illuminated by the entire source. Thus image uniformity can be obtained regardless of the source shape. (An extended coherent or partially coherent source is not useful, since the light from its various points would intefere to give nonuniformity in the object plane.) The effective source is the source image. For a uniform circular pupil and a uniform centered circular source, the ratio of the source image radius to that of the pupil, the *filing ratio*, is identical to σ discussed above. Coherent imaging systems, in which the source is a monochromatic point, use Koehler illumination (Tichenor and Goodman[51]). Condenser aberrations can reduce illumination uniformity. Another result of condenser aberrations is that the source image varies with the position in the field (Goodman[52]). Hence the image of a given pattern varies with its position in the field. If the source is not centered, object or image plane defocus results in lateral shifts of the image. (This can be easily observed with a microscope.)

In *critical illumination* (also called 'Nelsonian" in older literature) the source is imaged onto the object. (Fig. 33). Critical illumination is limited to small fields, since it requires a uniform source, such as a tungsten ribbon filament. For narrow fields, a capillary arc can be used. For shift invariance, the condenser exit pupil must be imaged into the entrance pupil of the imaging lens (which may require a field lens preceding the object). In

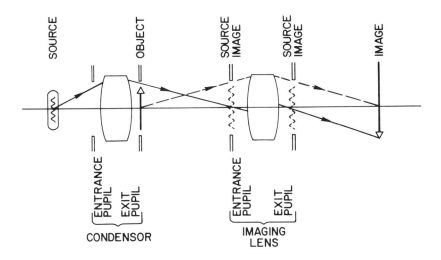

FIG. 32. A projector with Koehler illumination. The source is imaged into the entrance pupil of the imaging system, and the object is at the exit pupil of the illumination system.

practice, this requirement can be ignored if the field of view is small compared to the entrance pupil, as is the case with high-magnification microscope objectives. The effective source is the condenser exit pupil. For an axially symmetric projector, σ is the ratio of condenser numerical aperture to that of the imaging system. Zernike[53] showed that condenser aberrations are unimportant in critical illumination. (Zernike's paper contains an implicit assumption of stationarity.)

For *diffuse illumination* (Fig. 34), a good scatterer such as opal glass precedes the object, giving nearly noncoherent illumination, provided that the light from the source is sufficiently temporally noncoherent so that the diffuser produces no speckle. No phase match with the imaging system is required; so for shift invariance, the illumination system preceding the

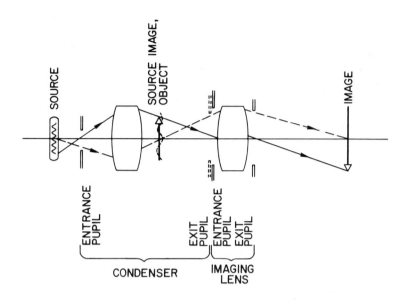

Fig. 33. A projector with stationary critical illumination. The source is imaged in the object plane, and the exit pupil of the illumination system corresponds with the entrance pupil of the imaging lens.

diffuser must only provide uniform irradiance. If ground glass is used as a diffuser, it must be sufficiently separated from the object plane, since the field just beyond ground glass is quite nonuniform. An integrating sphere provides diffuse and very uniform illumination. Diffuse illumination is

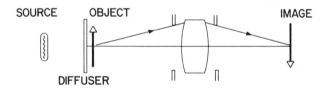

Fig. 34. A projector with diffuse illumination, which approximates noncoherent illumination if the diffuser is highly scattering and the light is not too monochromatic.

inefficient, which makes it impractical for some applications, e.g., photolithography. Photographic materials are sufficiently sensitive, so that it is often used in photographic copying.

If the image is recorded with a time-integrating detector obeying the reciprocity law, e.g., film, an extended source can be simulated by sweeping the region it would occupy with a point source, such as a focused laser beam (Lacombat et al.[54]). Critical illumination cannot be easily obtained in this way, since it would require near-perfect blending of the raster lines of the moving spot. Koehler illumination does not suffer in this manner, since the exact source shape is unimportant. The principal requirement of a scanning-source Koehler system is that for all source positions the object plane illumination be uniform. Any physically possible illumination mutual coherence function can be simulated with a moving spot.

Image power is often an important consideration. As the size of the effective source increases, a greater fraction of the incident light is diffracted by the object beyond the imaging system, so that for a source of fixed total power, coherence is decreased at the price of image energy loss. If the effective source overfills the pupil, energy is lost even with no object. More commonly, source brightness is given, in which case throughput is proportional to the area of the effective source. In a Koehler system, if the source size is given, then image power is increased by magnifying the source or, equivalently, by the condenser having as large as possible a numerical aperture on the source side. Thus, increased power is often accompanied by decreased illumination coherence. Another practical consideration is that of stray light, which may be abundant if the illumination fills the imaging lens, resulting in scattering from the edges of the lenses. In practice, the available light sources impose considerable restraints on illumination selection.

There is a body of folklore, erroneous generalizations, and much confusion having to do with illumination and partially coherent imaging. Few generalizations can be made with respect to imaging, since there are so many types of objects and so many parameters of the illumination and imaging systems.

One misunderstanding has to do with temporal and spatial coherence. Temporal coherence is not of primary importance in image formation, because the optical path lengths from each object point to its image are nearly equal. (Aberrations are tantamount to path length inequality.) The path lengths for different pairs of points are not equal, except for afocal systems. Thus a quasi-monochromatic source with, say, a 100 Å bandwidth gives virtually the same image as a source with a 10 Å bandwidth. However, temporal coherence does affect the interference noise due to scattering by imperfections, dust particles, etc.

There seems to be some confusion about the meaning of σ, probably because it is defined differently for Koehler and for critical illuminations. It is often stated that σ is the ratio of illumination to imaging numerical apertures, which is true for critical illumination, which is rarely used, but false for Koehler illumination.

It is sometimes stated that imaging is noncoherent or "effectively noncoherent" if $\sigma \geq 1$. As Figs. 31 and 32 suggest, this is approximately true for objects that vary only in amplitude transmittance, but untrue if there are phase discontinuities.

There is some confusion on the dependence of resolution on source size. This seems to arise because there are a number of resolution criteria and because the term "resolution" is used in many ways. In addition, a distinction is usually not made between microscopy, where the object is to be determined, and lithography, where an image is to be created. A rule of thumb in microscopy is that as coherence decreases, resolution increases. With decreased coherence, there are larger illumination angles, so light due to greater object spatial frequencies is diffracted into the imaging system and participates in the image formation, permitting finer object structure to be seen. On the other hand, the situation changes for the observation of phase objects, for which image contrast decreases with σ for the very same reason, namely that more Fourier components are accepted by the lens. In lithography, an object is not investigated. The aim is to produce a particular image with good contrast and steep edge slopes. The maximum spatial frequency of the image irradiance is determined by the image-side numerical aperture of the imaging lens, regardless of illumination.

A commonly held belief is that depth of focus increases as coherence increases. There is some truth here, but again it is an oversimplification. With coherent illumination, a defocused image can be quite different from the focussed image, but its contrast can be excellent since the fields forming the image can destructively interfere completely. In fact, with no a priori knowledge, it is often unclear when a coherently illuminated object is in focus. (This is easily demonstrable by imaging a piece of ground glass illuminated with a laser.) With noncoherent illumination, the image becomes monotonically fuzzier with defocus, but its form does not otherwise change. (Periodic objects are an exception, since their image modulation may decrease and then increase, going through several such cycles with diminishing maximum contrast.) Thus a technique for setting a lens for best focus is to temporarily decrease the illumination coherence. However, the situation is further complicated, since images of phase objects may increase in contrast with defocus.

The function $\tilde{C}(\mathbf{f}_1; \mathbf{f}_2)$ is often confused with the optical transfer function (OTF) or its magnitude, the modulation transfer function (MTF). The OTF is only applicable to noncoherent imaging, where there is a linear relationship between object and image irradiances. In partially coherent imaging, an "effective transfer function" can be defined as the ratio of the Fourier transforms of the image and object irradiances, $\propto \tilde{I}(\mathbf{f})/|O(\mathbf{f})|^2$. But this function is of doubtful value, since it is object-dependent and ill-behaved; it may, for instance, be infinite.

Projectors are used in metrological work. In workshop comparators,[33] for example, the silhouette of an object is imaged onto a screen, where it can be measured by translating a calibrated stage or be compared with a template. Low distortion is important in such applications. The principle of telecentricity (Fig. 35) is used so that defocus does not effectively introduce distortion. If the aperture stop is located at the rear focal plane of the portion of the imaging system preceding it, the entrance pupil is at infinity, so the principal rays and image forming bundles in object space

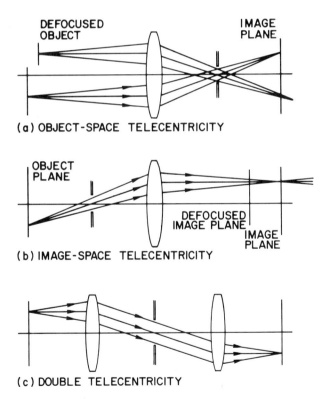

FIG. 35. Telecentricity. The ray bundles may be parallel to the lens axis in (a) object space, (b) image space or (c) both. Defocus in the telecentric space results in blurring, but not distortion.

are parallel to the lens axis. Thus, if the detector plane is fixed and the objective shifts axially, the image is blurred, but the "center of gravity" of the image of each point is unchanged. Similarly, if a system is telecentric in image space, magnification does not change with image plane defocus. Only an afocal system can be telecentric in both spaces, but it cannot be telecentric in just one.

In very precise metrology, such as that done on semiconductors, the effects of illumination on the images of edges and fine lines must be taken into account, as must the three-dimensional nature of the objects. In this regime, the scalar wave approximation is no longer adequate; imaging must be treated using electromagnetic theory.

Projectors have become particularly important in photolithography. In microcircuit production,[27] projectors are pressed to their limits by the requirements of resolution, repeatability, and overlay. For production, short exposure times are critical. Photoresist sensitivity is low and often limited to the blue and ultraviolet. With refractive optics, the required aberration correction can be obtained only over a narrow spectral region. Thus, powerful sources are required, with mercury arcs being the most common because of their strong spectral lines in the appropriate wavelength region. A recent development is the use of excimer lasers. Illumination must be uniform to a few percent across the mask, so that the size of the structures does not vary. Exact magnification setting and low distortion are critical, since solid state devices consist of a number of superimposed layers that may be exposed on different projectors. Photolithographic projectors are usually telecentric in image space. Both reduction (from 4× to 20×) and unit magnification systems (such as that shown in Fig. 21) are used. Photolithographic projectors are parts of complex systems that not only image, but align and focus the wafers.

4.7. The Microscope

Many aspects of the microscope are treated by Benford and Rosenberger,[55] Burrells,[56] Needham,[57] Chamot and Mason,[58] Francon,[59] Claussen,[60] Gray,[61] and Meyer-Arendt.[62] Microscope objectives are discussed by Benford[63] and Benford and Rosenberger,[64] who have compiled a list of microscope objectives and eyepieces. Delly and Wills[65] have compiled an extensive list of objectives. Meyer-Arendt[66] gives a large bibliography. Photoelectric setting microscopes, used in precision measurement, are discussed by Ciddor.[67] Much information is also available from microscope manufacturers.

When used visually, the microscope can be thought of as a sophisticated magnifier. The basic arrangement of the microscope imaging system is shown schematically in Fig. 36, where the objective and eyepiece are represented by simple lenses. The objective lens forms a real, enlarged image of the object, which is viewed through the eyepiece. Since the imaging is done in two steps, the term *compound microscope* is sometimes used. (The first microscopes of van Leeuwehoke had a single lens.) The first image is inverted and the second is not, and so the final image is inverted. The net visual magnification is the product of the objective transverse magnification m_{obj} and the eyepiece visual magnification m_{eye},

$$m_v = m_{obj} m_{eye}. \qquad (4.19)$$

Equations (4.3) and (4.4) for the angle subtended by the final image and the required magnification are applicable to the microscope.

In *photomicrography*, photography through a microscope the final image is real. Its size is related to that of the object by a transverse magnification, whose numerical value is not necessarily the same as that of the visual magnification obtained with the same instrument. The magnification can be found empirically by photographing an object with known dimensions, such as Ronchi ruling, or a *stage micrometer*, a precise glass rule. The measurement of magnification is discussed by Freeman.[68] Photographic magnification can be thought of as equivalent to a visual magnification if the photograph is viewed at the near point of the eye, in which case the angular subtense of the photographic image is given by Eq. (4.3), with m_v replaced by the transverse magnification of the photograph. The same comments apply to images viewed on a TV monitor.

Microphotography is the production of small pictures, for example on microfilm. It is discussed by Stevens.[69] The same principles are important in microlithography.

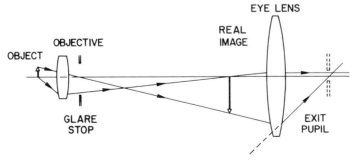

FIG. 36. The basic microscope imaging system, in which a real image is formed by the objective and viewed with the eyepiece.

160 BASIC OPTICAL INSTRUMENTS

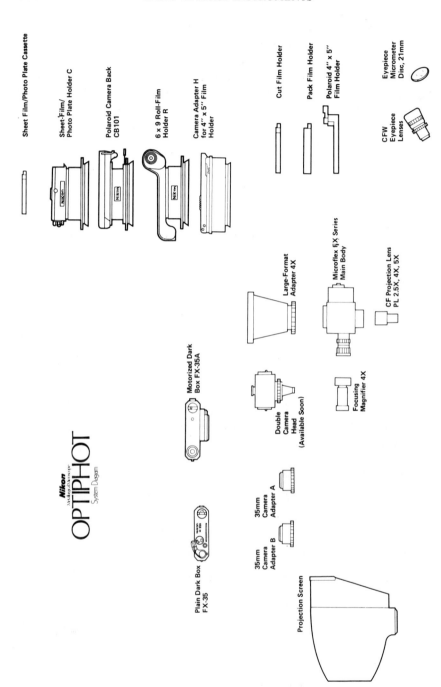

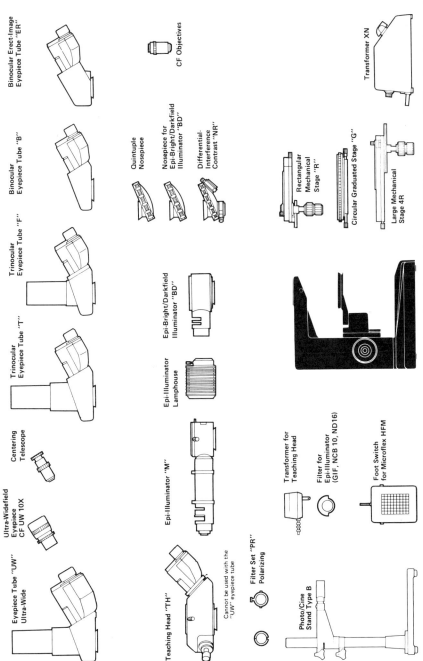

FIG. 37. A general purpose metallurgical microscope, showing the variety of attachments that are available. (Courtesy of Nikon Inc., Instrument Division, Garden City, N.Y.)

Modern general-purpose microscopes can be configured in a number of ways; and Fig. 37 shows the range of accessories available for a particular metallurgical microscope. Other microscopes are constructed for specialized tasks. Of the numerous variations of the microscope, some have only to do with the eyepiece, the objective, or the condenser individually. Others involve combinations of these, most commonly the objective and illumination systems. Many of these arrangements arise from a need to see the object other than simply enlarged. The image is modified or "processed" in order to enhance, to make visible, or to measure some feature of interest. The principles of these schemes are not limited to microscopes.

A number of physical parameters of microscopes, which are discussed below, are shown in Fig. 38. Many are not standardized, so interchanging

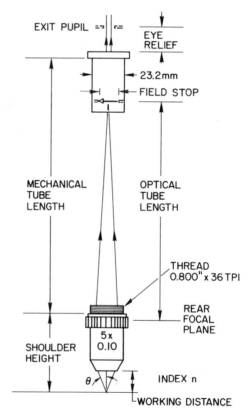

Fig. 38. Microscope physical parameters and typical dimensions.

TABLE I. Microscope Objective Nominal Characteristics

Ideal properties of a series of microscope objectives[1]

magnification, m_{obj}	5×	10×	20×	50×	100×
numerical aperture, NA	0.09	0.20	0.40	0.85	0.95
resolution[2] [μm]	3.7	1.7	0.84	0.39	0.37
depth of field[3] [±μm]	34	6.9	1.7	0.38	0.34
cut off frequency[4] v_{max} [lines/mm]	330	730	1450	3090	3450
cut off period[4] p_{min} [μm]	3.0	1.4	0.69	0.32	0.29
field diameter[5] [mm]	3.6	1.8	0.9	0.36	0.18
resolvable lines per field[6]	3500	3900	3900	3300	1900
exit pupil diameter[7] [mm]	0.9	0.95	1.0	0.85	0.45
working distance[8] [mm]	12	17	0.75	0.20	0.10

[1] The magnifications, numerical apertures, and working distances are those for a "typical" set of commercially available microscope objectives. The resolution and depth of field have been computed for $\lambda = 0.55$ μm, assuming aberration correction. The number of resolvable lines in the field is given under the assumption of aberration correction over the entire field of view. In fact, objectives are best corrected on axis and are not free of aberrations everywhere in the field.
[2] Rayleigh criterion for noncoherent imaging, Eq. (4.22).
[3] Quarter wave criterion, Eq. (4.24).
[4] Eq. (4.23).
[5] Used with an eyepiece with a 18 mm diameter field stop.
[6] At the spatial frequency of 60%, modulation for noncoherent imaging of a black and white periodic object with equal lines and spaces.
[7] With a 10× eyepiece.
[8] There are no typical values for working distance, which depends on the intended purpose of an objective and on details of its design.

parts may degrade performance. Table I gives a number of parameters, treated below, having to do with the performance of a series of objectives. The transverse magnification at which an objective is intended to be used is usually referred to as its *power* or *magnifying power*. This number is usually engraved on the objective, where it is indicated by a times sign, for instance 10×, or ×10 for a transverse magnification of −10. As with any lens, a microscope objective can be used at any magnification, but its aberrations are best corrected at only a single magnification. Thus, in speaking of an objective's magnification, object and image positions, and so on, we mean those for which it was designed. Aberration correction also takes into account the thickness and refractive index of cover glasses, which are commonly used in biology. (Cover-glass thickness is typically 0.17 mm.) The effect of a cover glass on aberrations increases with numerical aperture, and is of little importance for numerical apertures up to 0.3 or 0.4. Fluid immersion also influences aberration correction. Objectives may be quite

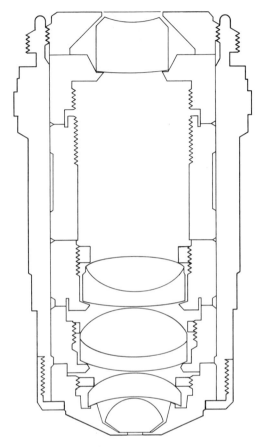

FIG. 39. A high magnification, high-NA microscope objective shown in cutaway. (Courtesy of American Optical.)

complicated, as the cross-sectional view in Fig. 39 shows. In general, the complexity increases with magnification and numerical aperture, as well as with color correction and field flatness.

Microscope objectives (as well as other lenses) are often named according to their degree of aberration correction. *Achromatic* objectives are corrected for color at two wavelengths, a minimum level of correction, unless the instrument is used monochromatically. *Apochromatic* objectives are further corrected at a third wavelength, which requires fluorite or special glasses. *Aplanatic* objectives are free from spherical aberration and coma. *Plano* objectives have a flattened field, which is more important for photography than for visual work. These terms are not rigorously standardized. They serve mainly for comparison within a given manufacturer's line. The objective is usually located axially by a shoulder. The predominant thread size is 0.800 inch diameter and 36 threads per inch, which derives from the nineteenth-century standard of the Royal Microscopical Society (RMS). Other means of mounting, such as devetail slides, are occasionally used.

The *shoulder height, parfocal distance*, or *adjustment distance* is that from the object plane to the locating shoulder. Several manufacturers have recently begun to standarize this distance at 45 mm. The *working distance*, or *free working distance*, is the separation between the object and the first surface of the objective. For biological objectives, this distance may be measured to the top of the cover glass. In general, as Table I shows, working distance decreases as magnification and numerical aperture increase. The working distance may be less than a millimeter, so objectives often have spring-loaded front ends to avoid damaging collisions.

The *optical tube length* is the distance from the rear focal plane of an objective to the real image position. For all but the lowest power objectives, the tube lengths for a set of objectives are nearly equal. The tube length 160 mm, often quoted in texts, is common, but is not universal. Other common lengths are 170 mm and 215 mm. The focal length and magnification are related to the optical tube length by Newton's relation [Eq. (2.45)], which gives

$$\text{optical tube length} = f_{\text{obj}} \ |m_{\text{obj}}|. \quad (4.20)$$

The *mechanical tube length* is the physical length of the tube that connects the objective and eyepiece. The eyepiece usually slips into the tube and is located by a shoulder. The distance from the locating shoulder to the object of the eyepiece is not standardized.

The term "tube length" may be used to mean the optical tube length, the mechanical tube length, or the distance between the objective locating shoulder and the image position. Different objectives for which both the shoulder-to-image distance and the shoulder height are identical may be interchanged (as by a rotating turret) without refocusing the eyepiece. Such objectives are called *parfocal*, and objectives with different magnifications are available in parfocal sets. A fine adjustment for parfocality is sometimes made by spacers above the objectives. Parfocality cannot be obtained if the axial position of the eyepiece is incorrect. A set of objectives on a turret that are mutually centered is said to be *parcentered*.

Infinity-corrected objectives work with the object at their front focal plane and its image at infinity, so there is no specific length (Fig. 40). An auxiliary lens, sometimes called a *tube lens*, acts as a telescope objective, forming a

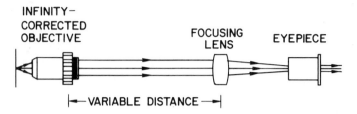

FIG. 40. The arrangement of a microscope with an infinity-corrected objective. The object-to-image distance is not fixed. The objective and focusing lens form a real image that is viewed through the eyepiece. Alternately, the combination of the focusing lens and eyepiece can be thought of as a Keplerian telescope, viewing an image at infinity formed by the objective.

real image viewed with the eyepiece. The magnification of the objective and tube lens pair equals the ratio of their focal lengths. So the magnification inscribed on an infinity-corrected objective implies that it is used with a certain standard tube lens. This arrangement permits easier parfocality adjustment and simplifies modular construction. Vignetting sets a practical upper limit on the objective–tube lens separation.

Microscope objectives, except those of the lowest power, usually have a stop near their locating shoulder. The stop diameter is typically between 7 mm and 9 mm. For many objectives this is the aperture stop, which may be located at the rear focal point of the objective, which is, thereby, telecentric in object space.

The numerical aperture, or NA, of an objective (or of any other lens) is

$$NA = n \sin \theta, \quad (4.21)$$

where θ is the half angle of the axial cone of light accepted by the objective, and n is the index of refraction in the object space. For non-infinity-corrected objectives, the numerical aperture on the image side equals that on the object side divided by the magnification. Objectives are usually inscribed with their nominal numerical apertures. There are devices, called *apertometers*, to measure NAs.

For an objective that is free from aberrations, i.e. "diffraction limited," its numerical aperture and the wavelength determine the limitations of imaging having to do with the wave nature of light, especially resolution and depth of field. With respect to resolution, the fundamental length is λ/NA, where λ is the vacuum wavelength. An estimate of the resolution is given by the Rayleigh criterion, according to which two noncoherent points of equal intensity can be regarded as distinct when their separation equals the radius of the first zero of their diffraction patterns (Figs. 24, 25,

26, Chapter 2 Volume 26). This separation is

$$d = 0.61 \frac{\lambda}{NA}. \quad (4.22)$$

This "limit" is a rule of thumb that is directly related to viewing an object consisting of two points, rather than a continuous object. (This rule of thumb is more applicable in astronomy.) Resolution also depends on signal-to-noise ratio and it is changed by nonlinear image processing, e.g., electronically or with film. Other limits of an objective are its cutoff spatial frequency and period,

$$v_{max} = \frac{2NA}{\lambda} = \frac{1.4}{d} \quad \text{and} \quad p_{min} = \frac{\lambda}{2NA} = 0.7\, d. \quad (4.23)$$

Object amplitude Fourier components with spatial frequencies greater than v_{max} or periods less than p_{min} do not contribute to image modulation, even with illumination angles as great as 90° from the lens axis. Periodic objects whose amplitude transmittances have periods less than p_{min} have uniform images, regardless of the method of illumination. Noncoherently illuminated periodic objects with equal lines and spaces and unit contrast, e.g., Ronchi rulings, have diffraction-limited image modulations of greater than about 60% nominal if their periods are greater than $3p_{min}$. (The above equations are written in terms of the object space numerical aperture and object distance. The expressions for the image space are formally identical, since the transverse magnification equals the ratio of the object to image space NAs.)

The *depth of field* is the maximum axial displacement of the object from the position of best focus that leaves the image sensibly unchanged. It is usually taken to be the displacement at which there is a quarter wave of defocus. (The $\lambda/4$ here refers to the separation at the rim of the aperture between a spherical wave originating from an object point in focus and one defocused.) If there are no aberrations, the depth of field is

$$\delta = \pm \frac{n\lambda}{2\, NA^2} = \pm \frac{\lambda}{2n \sin^2 \theta}. \quad (4.24)$$

Depth of field is changed by aberrations, which cause the image to change differently with defocus. Spherical aberration, for example, causes the image to be different on opposite sides of best focus, and with astigmatism the plane of best focus for a line depends on its orientation. Resolution and depth of field of a lens are related. Eliminating NA in Eq. (4.23) and Eq. (4.24) gives for an aberration-free lens, $\delta = 2np_{min}^2/\lambda \sim nd^2/\lambda$. For a given resolution, shorter wavelength gives greater depth of field, as does liquid immersion.

Objectives that work in air have numerical apertures up to about 0.95, corresponding to a half angle of 71°. NAs as great as about 1.40 are obtained by *immersion objectives*, in which a liquid fills the space between the object and the first objective surface. There are special immersion oils for this purpose and the index of the oil may match that of the first objective element or the cover slip, if used. The increase in NA with the index of refraction of the object space can be thought of as arising from a decrease in wavelength, giving smaller diffraction angles for a given object (Fig. 31, Chapter 6), so that more of the diffracted field is captured by an objective with a given collection angle. In general, objectives with greater magnification have greater numerical apertures, so that increased magnification is accompanied by increased resolution. Some objectives have adjustable apertures, and hence variable NAs.

Any geometrical magnification is possible, but just as a photographic image can be enlarged until there is no further revelation of detail, so a microscope can magnify uselessly. For example, if the first diffraction orders of a periodic object fall beyond the collection cone of the objective, then regardless of the magnification, the image is uniform, consisting only of the zeroth order. *Empty magnification* is that above which no further detail is revealed. For visual purposes, magnification is empty when the image of the distance just resolved by the microscope subtends an angle greater than that of the eye's acuity. To determine when this occurs, set s in Eq. (4.4) equal to the diffraction resolution distance of Eq. (4.22), giving for $\lambda = 0.55$ μm,

$$m_v \sim 230 \text{ NA}. \tag{4.25}$$

This value is somewhat arbitrary, since its derivation makes use of two imprecise quantities, the resolution of the eye and the resolution of the microscope. Accordingly, there are other estimates of the empty magnification condition, a common one being $m_v = 1000$ NA. These values are also applicable to photomicrography, if the photograph is viewed at the near point of the eye. Empty magnification can be approached in a different way. If the exit pupil of a visual instrument is smaller than that of the eye's pupil, then the eye is effectively stopped down. The image then appears blurred, since the eye's Airy disc extends over several rods or cones. The optical invariant [Eq. (2.27)] relates the objective numerical aperture and object field size to the exit pupil size and image field angle. If h is the radius of the object field, then $2h \text{NA} = d_e \sin \psi \sim d_e \psi$, where d_e is the exit pupil diameter and ψ is the angle of the chief ray in the exit pupil. Using Eq. (4.3) to eliminate h in favor of m_v gives

$$\text{exit pupil diameter} = \frac{\text{NA}}{m_v} \times 500 \text{ mm}. \tag{4.26}$$

With a 10× eyepiece, the exit pupil diameter in millimeters is $50\,\text{NA}/m_{\text{obj}}$. If this quantity is less than about 2 mm, there is empty magnification. The prediction of Eq. (4.25) corresponds to an exit pupil of 2 mm. The value $m_v = 1000\,\text{NA}$ corresponds to an exit pupil of 0.5 mm.

As Table I shows, objectives typically have $m_{\text{obj}} \sim 50\,\text{NA}$, and exit pupil diameters are typically about 1 mm, except at the highest magnifications, where numerical aperture cannot be correspondingly increased. Likewise, the nominal number of resolvable lines per field is nearly constant for the set of objectives, except at the highest magnifications. Different combinations of objectives and eyepieces can be used to obtain the same overall magnification, according to Eq. (4.19). In general, lower NA objectives provide less resolution but longer working distance and greater depth of field.

There are several special types of objectives. *Long working distance* objectives may be required, for example, for clearance to manipulate an object or to view an object within a chamber. *Heat-resistant objectives* are used to view heated objects, for instance to measure melting points. (Heating and cooling stages for microscopes are also available.) For radioactive objects, there are objectives with special glasses that do not become discolored by radiation.

Reflective, or catoptric, objectives[59,70] have a larger working distance than refractive objectives for a given numerical aperture. Reflecting objectives are based on the arrangement of Fig. 41, similar to the objective of a Cassegrain telescope. Some objectives derived from this form are catadioptric, that is, they contain both reflecting and refracting elements. Reflecting objectives have no chromatic aberration, so they can be focused in the visible to obtain images in the infrared (IR) or ultraviolet (UV). (Note,

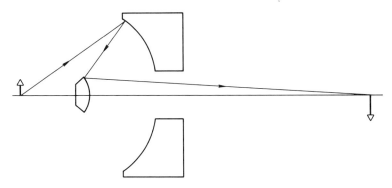

FIG. 41. A reflective microscope objective. Such objectives are completely achromatic and can provide high numerical aperture and long working distance. However, their aberrations can be corrected only over small fields.

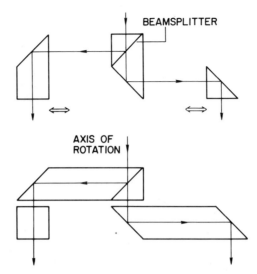

FIG. 42. Two binocular viewing devices with variable interpapillary distance. The optical path length must not change with eyepiece separation. In the upper arrangement, as the eyepieces separate, they also move toward the objective.

however, that the focal depth is less in the UV.) These objectives have central obstructions, so standard illuminators are not ideal for them. The aberrations of reflecting objectives can be corrected only over a small field.

The discussion of eyepieces above applies to those for microscopes. The most common eyepiece barrel diameter is 23.2 mm, and the diameter of the eyepiece field is typically about 18 mm. A wider field requires a tube of larger diameter. Eyepieces are usually engraved with their visual magnifications, which range from 5× to 25×, with 10× being the most common.

In the *binocular* microscope, identical images are presented to the eyes by an apparatus such as one of those shown in Fig. 42. Adjustment for eye separation (interpupillary distance) is made by a mechanical translation or rotation. The binocular microscope differs from the stereo microscope (not discussed here), in which the two eyes view the object at different angles. The *trinocular* head allows binocular viewing and provides a third image for photography or another application.

4.7.1. Illumination Systems

Microscope illumination systems of the Koehler and critical types have been used, but the latter is now rare. A transmitted-light Koehler system is shown in Fig. 43. The source is imaged into the front focal plane of the substage condenser lens and reimaged into the stop near the rear focal

THE MICROSCOPE 171

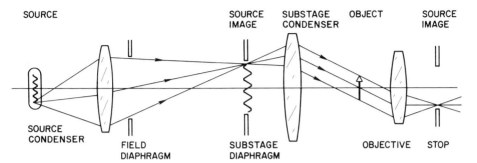

FIG. 43. Koehler illumination typically found in microscopes. The source diaphragm is imaged into the objective stop, and the field diaphragm is imaged in the object plane. If the objective is telecentric in the object space, then each source point illuminates the object with a plane wave.

plane of the objective. A *field stop* conjugate to the object plane limits the illuminated region, which should not much exceed the field of view in order to reduce glare. An aperture stop, or *substage condenser diaphragm*, delimits the source image, thus controlling the spatial coherence of the illumination. This stop should not be used to change the illumination level, since it also changes the form of the image. Condensers often have a flip-in lens immediately below the stage, to increase their numerical aperture for use with high NA objectives. Thus, light is not wasted by overfilling low NA objectives. For very large NA immersion objectives, there are liquid immersion condensers.

Condensers have centering and focus adjustments. If the illumination system is properly set up, its field and aperture stop are focused and centered in the appropriate planes. To check this, close the field stop and observe its image in the object plane, where it should be focused and centered. In addition, remove an eyepiece and look in the tube at the image of the condenser diaphragm, which should be centered and focused at the rear stop of the objective. A device called the phase telescope can be used to better view this region.

Opaque objects must be illuminated from the side of the imaging system. The arrangements that do so are called *vertical illumination, incident light*, and *episcopic* or *epi*. The term "vertical illumination" is usually applied to illumination through the objective (Fig. 44), regardless of the actual orientation of the microscope. The other terms are often used in a more general way, including vertical as well as oblique illumination. Microscopes with vertical illuminators are often called metallurgical, as opposed to those that work in transmitted light, which may be called biological. In illuminating through the objective, scattered light is a problem. A good light trap is

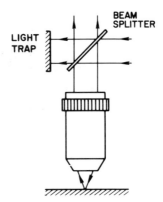

FIG. 44. Vertical illumination, such as that common in metallurgical microscopes. The light trap is particularly important, since light that is not absorbed is reflected back into the system, resulting in loss of contrast. Many objects scatter less light back into the system than does a poor light trap.

required to absorb the light from the illuminator that is transmitted by the beam splitter. The amount of light scattered by a bad trap may easily exceed the light reflected by an object into the objective. To check the adjustment of a vertical illuminator, use a front surface mirror as the object, and follow the procedure given above.

In *dark-ground* or *dark-field illumination*, the illumination is directed beyond the objective, so that only light that has been diffracted contributes to the image (Fig. 45). Thus flat, homogeneous transparent objects and

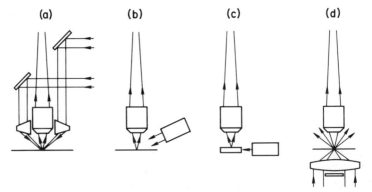

FIG. 45. Four varieties of dark-field illumination. Only light scattered by the object enters the objective. Types (a) and (b) are used for opaque objects. Most commercially available dark-field objectives use the principle shown in (a). Illumination (c) is used for particles within a liquid or imperfections in a clear solid. This arrangement is referred to as ultramicroscopy. Arrangement (d) is used with transmitting objects.

planar, specular opaque objects give dark images, and discontinuities in otherwise uniform objects have bright images. For example, edges on an otherwise flat and smooth surface show up as bright lines. Dark-field systems can be used to detect subresolution particles, for example, dust on a smooth surface. In what is called *ultramicroscopy*, suspended particles within a transparent volume are investigated with illumination directed perpendicular to the axis of the imaging optics. The sizes of subresolution particles can only be inferred from the amount of light they scatter. The illumination modes can be combined in a number of ways. For example, *Rheinberg illumination*[71] is a combination of bright-field illumination in one color and dark-field in another.

4.7.2. Special microscope apparatus

Phase contrast and *interferometric* apparatus involve both the illumination and imaging portions of the microscope, which must be well aligned. Phase contrast methods (Chapter 6) are used in transmission to see clear objects, whose refractive indices differ from their surroundings, for instance cells. In reflection, these techniques enhance small variations in height. There are a number of phase contrast methods, many of which are described by Bennet et al.[72] and Francon.[59] There is a number of interferometric attachments for microscopes, some of which are not much used outside of microscopy, e.g., the Nomarski and Mireau. The eyepiece can wholly contain a shearing interferometer.[59] For more information, see Tolansky,[73] Francon,[59] and Krug et al.[74]

Many general-purpose microscopes have polarization accessories, but the full-fledged *polarization microscope* is a specialized instrument. The objectives have a centering adjustment, and they are stress-free to eliminate birefringence, which would introduce spurious effects. Variable azimuth polarizers, retarders, and the like can be placed in the imaging and illumination systems, and there is a rotating stage. Polarizing microscopes can be used in two ways. In the *orthoscopic* mode, which requires a stopped-down condenser, the illumination is nearly collimated and parallel to the microscope axis. In the *conoscopic* mode, the numerical aperture of the illumination is high, and the object is not viewed. Instead, the rear focal plane of the objective is observed by means of a lens in the tube, called a *Bertrand lens*, that images this plane at the eyepiece. The illumination can be thought of as consisting of plane waves, which are focused at the rear focal plane of the objective, where each point corresponds to a direction through the object. The crystalline structure of the sample determines the observed pattern. To quickly change the mode of operation, the Bertrand lens usually flips in and out, as does a lens in the condenser that changes the illumination

NA. Polarization microscopes are discussed by Hartshorne and Stuart[75] and in many books on optical mineralogy.

In *fluorescence microscopy*, the object is illuminated with blue or ultraviolet light, and the visible fluorescence is viewed.[76] If the excitation light is visible, then filters that pass longer wavelengths are used in the imaging system to pass only the light produced by the fluorescence. Dark-field illumination can also be used to avoid seeing the excitation light. The glass in the objective must be nonfluorescent, so that no background light is produced. Fluorescence techniques are used, for example, with special stains in biological work.

In *infrared microscopes* and *ultraviolet microscopes*, the real image is converted to the visible. There are image converters that replace the eyepiece of standard microscopes, extending them into the near infrared, where many standard dioptric objectives still give satisfactory images. Ultraviolet microscopes require quartz, fluorite, or all reflective objectives. One application of such microscopes is seeing through objects that are opaque in the visible. In the near IR, for example, circuits beneath the top layer of a silicon chip can be observed.

In a *scanning microscope*,[77,78,79,80] the "image" is the record of a detector output obtained when a spot of light is moved across the object. Such instruments often have projector analogues (Fig. 46). A projector and a scanner with identical optics give the same images if the projector source is replaced by a detector of the same shape, whose response is linear with incident power, and if the scanning is done by a subresolution point in what would be the projector image plane. In the original "flying spot microscope" of Roberts and Young,[81] a cathode ray tube (CRT) was used

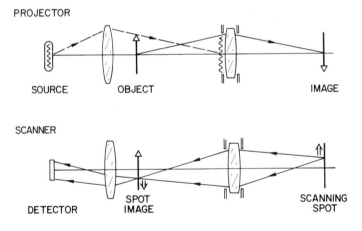

FIG. 46. A scanning microscope and its equivalent projector.

to provide a moving spot, but now focused laser spots are usually used.[82] The CRT has the advantage of speed and random access, but is far less intense than a laser. In another type of scanner, the spot of light is fixed and the object translated, which gives an unlimited field of view with a focusing lens that is aberration-corrected only on axis. For these scanners, the same equivalence does not hold in general, but there is often a different equivalent projector. The various possibilities are discussed by Goodman.[43] In the *confocal* scanning microscope,[80] the detector is preceded by a pinhole that is conjugate to the scanning spot. This configuration gives better resolution and smaller depth of field than the others, at the cost of less energy and more difficult alignment. A unique property of the confocal microscope is that the image intensity drops with defocus, so a single slice of the object can often be isolated. The video disc read-out head (Bouwhuis et al.[83,84], Isailovic[85]) is an example of a moving object scanner. Such a device must also be capable of automatic focusing and tracking.

The *microdensitometer*,[86,87] another type of scanner, is used to map the optical density of photographic plates, the opacity of chrome-on-glass masks, etc. These are measurements of irradiance transmittance, and it is usually desirable to ignore phase variations due to object relief. This means that the equivalent projector is illuminated noncoherently. Microdensitometers are said to be linear insofar as they are insensitive to object phase, and one of their specifications is the spatial frequency beyond which linearity breaks down. For a given imaging system, linearity is extended by using detector optics of the largest possible NA. The imaging abilities of microdensitometers are often specified by an MTF, which is applicable only in the linear regime.

4.8. Schlieren Instruments

Schlieren systems show variations in the optical path length of light passing through transparent objects or reflecting from the surface of specular objects by causing the path length differences to appear as irradiance variations. The German word *Schlieren*, meaning streaks or striations, comes from their appearance. In the widest use of the term, schlieren systems include interferometers and phase contrast instruments, which are discussed in Chapters 5 and 6. We restrict ourselves here to those instruments that are directly sensitive to optical path gradients.

The first schlieren method was the *Foucault knife-edge test*,[88] invented in 1858 for testing optical components. Toepler[89] employed this method for

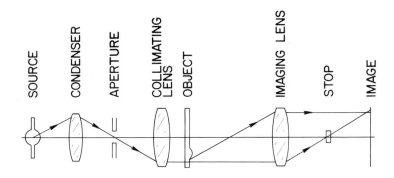

FIG. 47. A basic schlieren system. This a dark-field instrument, in which object heterogeneities appear as bright regions.

the study of gaseous inhomogeneities. Two of the most extensive discussions are those of Vasil'ev[90] and Wolter.[91] Merzkirch[92] gives a survey of techniques, and Ojeda-Castañeda[93] discusses the knife-edge test and its relatives. See also North,[94] Beams,[95] and Holder and North.[96] Applications of schlieren methods include the testing of optical components and fluid studies, for example wind tunnel tests and convection analysis. Index variations as small as 10^{-6} can be detected; so 1° C is the detectable temperature variation for air at standard temperature and pressure.

Most schlieren systems can be thought of as variations of the Koehler illuminated projector shown in Fig. 47, which is a dark-field system. Here light from a small source is imaged by the condenser onto a stop that sharply delimits its image. The light is collimated and passes through the object under investigation, here a plate that is nominally plane and parallel. The object is imaged by a second lens that also forms an image of the source at an opaque stop. If the object is transparent and homogeneous and if its surfaces are flat and parallel, a good source image is formed, which is blocked by the stop, giving a uniformly dark image. But if the plate is imperfect, the directions of the light rays are changed, so the rays miss the stop and give bright regions in the image. The amount of ray bending depends upon the gradient of the optical path length through the object in the plane perpendicular to the system axis. From the Fourier optics viewpoint, the stop is a spatial filter that removes the zeroth order, giving a dark field when the object is homogeneous. Inhomogeneities introduce higher spatial frequencies, and hence bright regions. In practice, the size of the source and stop are finite, so a range of the low frequencies is also reduced, if not removed.

The requirements on the optics for high-sensitivity schlieren systems can be understood from Fig. 47. The source should be bright, since only a small portion of the light contributes to the image, and small, for maximum sensitivity to gradual object variations. Scattered light should be minimized, since a background level in the final image plane reduces sensitivity. The quality of the condenser in a schlieren system is more critical than is usually the case. The condenser optics must be adequately corrected for aberrations, since these are indistinguishable from object variations. With no object in place, a uniform image field is a necessary condition for proper performance.

A variety of schlieren instruments has been devised. For greater sensitivity, there are multiple-pass systems, in which light traverses the object several times to increase the effect of optical path length variations. Object gradients can be mapped with a small aperture in the source image plane. For each aperture position, a small range of object spatial frequencies contributes to the image, and those portions of the image are bright for which the object has a particular gradient. Colored filters can be used to determine where the light crosses the stop plane, by producing image color indicating the direction and magnitude of the gradient. If gradients are merely to be detected, such systems take advantage of the eye's sensitivity to color variations, which is greater than that to intensity variations.

In a bright-field complement of the instrument of Fig. 47, the stop is replaced by an aperture that just passes the nominal image of the source. In this case, object inhomogeneities give rise to dark image regions. If the object is spatially band-limited, with a maximum frequency within the cutoff frequency of the imaging system, the image irradiance is complementary to that obtained with the dark-field system.

Thick objects should be studied in collimated light, which is provided by the "Z-system" (Fig. 48), used for gas dynamics studies. For thin objects collimation is not necessary, and the stop can be placed before the imaging lens to reduce the importance of scattering by this lens (Fig. 49).

In the classical Foucault knife-edge test (Fig. 50), a nominally spherical mirror reimages a point source near its center of curvature. If the imaging is aberration-free, the entire mirror appears to change quickly from light

178 BASIC OPTICAL INSTRUMENTS

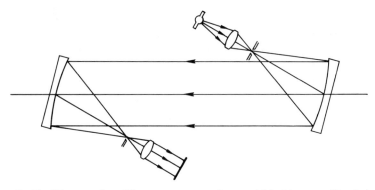

FIG. 48. The "Z-system," a schlieren arrangement for very thick objects, used in wind tunnel studies.

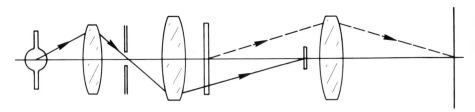

FIG. 49. A thin-object schlieren system. The illumination converges, so the stop can be located before the imaging lens to reduce the effect of its scattering.

to dark as the knife-edge passes through the source image. This is not so if there are aberrations, each of which gives rise to a characteristic pattern. Further information about the mirror can be obtained by moving the knife-edge axially. Linfoot[98] treats the theory of this test.

The knife-edge is used with other arrangements, for example, as the stop in both the illumination and the imaging systems, allowing fine adjustment

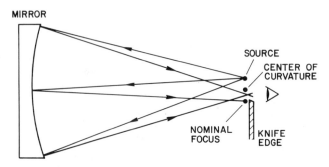

FIG. 50. The Foucault knife-edge test, shown here arranged for testing a concave mirror.

of the dark-field condition and high sensitivity (Fig. 47). From a Fourier optics viewpoint, the knife-edge in the imaging system acts as a step filter that shows only the tops of regions of greater path length or the bottoms of those of lesser length (or vice versa). There are a number of other tests analogous to the knife-edge that use obstructions of different shapes, e.g., single wires, rulings, and clear plates with phase discontinuities (Vasil'ev[90]).

For objects that vary in only one direction, some special arrangements can be used. Light flux can be increased without loss of sensitivity by using elongated sources and stops. Lloyd[99] and van Oss[100] describe several systems whose output irradiances show gradient versus position.

There are a number of arrangements analogous to those discussed above, but in which the imaging system contains a stop that is not totally opaque. It is partially transmitting and also shifts the phase of the light passing through it. The best known of these arrangements is the phase-contrast arrangement of Zernike.[101] Another is the *Lyot test*,[102] which shows fine surface roughness, but not slowly varying shape.

The so-called *shadow test* is sometimes counted as a schlieren method. The light from a small, bright source passes through the object to a screen, where irradiance nonuniformities arise from object inhomogeneities. This test does not show gradients directly. Its advantage is its simplicity, little optics being required. A zirconium arc is useful as the source for this test.

The distinction between the schlieren and dark-field technique may be blurred. Schlieren systems are usually set to detect fine object gradations, while dark-field systems detect sharper changes. An example of the latter are the optics used to observe bubble chambers, in which the bubbles are made to stand out as bright against a dark background (Welford[103]).

4.9. The Ellipsometer

The *ellipsometer* measures the polarization changes in light reflected from a surface. In the instrument shown schematically in Fig. 51, a collimated beam of known polarization is prepared and reflected from the surface under study. The outgoing beam is processed by a second polarization system. Its intensity is measured, and the change in polarization deduced.

An extensive treatment of ellipsometry is given by Azzam and Bashara,[104] and there are numerous papers on techniques and applications in the ellipsometry and polarization symposia proceedings (Passaglia et al.,[105] Bashara et al.,[106] Bashara and Azzam,[107] Muller et al.,[108] Hyde and Azzam,[109] Azzam and Coffeen[110]). Hall[111] and Rothen[112] discuss the history of ellipsometry, which can be traced to Malus's discovery that light can be polarized by reflection from a dielectric. Anomalies later encountered were found to

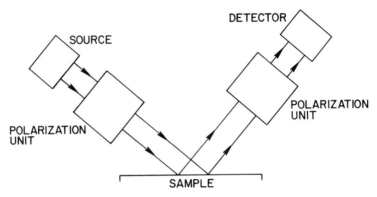

FIG. 51. The general arrangement of an ellipsometer.

be caused by thin surface layers. Film thicknesses and thickness differences of only a few angstroms are measurable.

Ellipsometry was originally concerned only with the specular reflection of completely polarized light. Hence its name, which comes from the fact that, in general, the electric and magnetic field vectors of completely polarized light describe an ellipse. In traditional ellipsometry, which was developed before the existence of photodetectors, the magnitude of the reflected light is not measured. Instead, nulling procedures are used, in which the ellipsometer components are adjusted so that the output vanishes, a method suitable for the eye and also for nonlinear photodetectors.

For such techniques, the required components are a polarizer following the source (unless its output is already polarized), an analyzer preceding the detector, and a compensator, in either arm, between polarizer and analyzer. Four parameters can be adjusted: the compensator retardation and the azimuths of the polarizer, analyzer, and compensator. There are six combinations of these, taken two at a time, for which nulling can be achieved. The azimuths are usually varied, since they can be measured more accurately (0.005°–0.01°) than the retardation (0.1°–1.0°). There are three methods for locating the null. It can be found with just two adjustments, but less accurately than with an iterative method that terminates when successive changes become too small to be measured. In automated nulling ellipsometers, the iterations are performed by stepping motors under computer control.

In the manual ellipsometer configuration shown in Fig. 52, light from the source is filtered, collimated, and elliptically polarized by a linear polarizer followed by a compensator. This arrangement is called a PCSA (polarizer-compensator-system-analyzer); in the PSCA configuration the compensator is on the output arm. The sample can be viewed for alignment and for

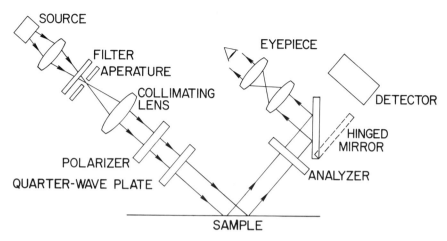

FIG. 52. A manual ellipsometer arrangement.

selecting the sample region of interest. The null can be approached visually, with the detector used for the final zeroing. A pair of general purpose commercial ellipsometers with vertical and horizontal sample orientations is shown in Fig. 53.

The study of partially depolarizing surfaces requires *Mueller matrix ellipsometers*, which determine the Stokes vector of the output beam. This cannot, in general, be done by null measurements; Hauge[113,114,115] lists the methods of making these measurements, most employing a periodically varying component. If a rotating element is used, the Stokes parameters of the light are linear combinations of the Fourier coefficients of the detector output. For example, in the arrangement due to Sekera,[116] the detector is directly preceded by a fixed analyzer, in front of which is a rotating 90° compensator. If θ is the angle between the fast axis of the compensator and the analyzer transmission azimuth, the power reaching the detector is

$$P(\theta) \propto (S_0 + \tfrac{1}{2}S_1) + \tfrac{1}{2}S_1 \cos(4\theta) + \tfrac{1}{2}S_2 \sin(4\theta) - S_3 \sin(2\theta), \quad (4.27)$$

where (S_0, S_1, S_2, S_3) is the Stokes vector of the beam. Since the analyzer is immediately before the detector, variations of detectivity with polarization are unimportant. To avoid mechanical motion, electrically oscillating elements, such as Faraday modulators and phase modulators, can be used, but the analysis of the output is more difficult. A method of determining polarization with four detectors and no changing elements has been devised by Azzam.[117]

Systems similar to that of Fig. 51, called *polar nephelometers*, are used to determine the scattering properties of volume samples, as well as surfaces.

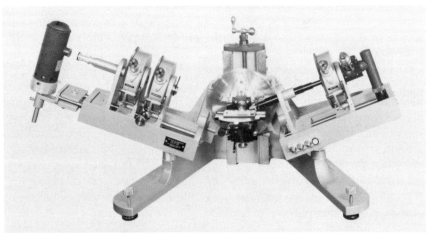

FIG. 53. Two general-purpose commercial ellipsometers. (Courtesy of Gaertner Scientific Corporation, Chicago, Ill., and Rudolph Research, Fairfield, N.J.)

This requires measuring the power and polarization of the light scattered at any angle for any input direction and polarization.

Traditional ellipsometers use a narrow spectral bandwidth, since the action of the compensator varies with wavelength. The more recent development of spectroscopic ellipsometry is discussed by Aspnes.[118]

4.10. Refractometers

Refractometry is the measurement of the index of refraction. Often only the real part of the index is measured. Refractometry is done with a variety of instruments, some performing only this task, and others that are more general. In principle, any instrument whose behavior depends upon the refractive index of one of its components—including what are usually empty spaces—can be used as a refractometer. Most refractometers work by either goniometric or interferometric means. The objects of refractometry may be solids, liquids, or gases, and may be transparent or absorbing. The index may be measured at a single standard wavelength, or over a range of wavelengths. Fishter,[119] Longhurst,[120] and Werner[121] discuss a number of refractometers. Some aspects of refractometry are treated in analytical chemistry books, for instance that of Gibb,[122] and others are discussed in crystallography and mineralogy books.

Refractometers may measure either the absolute or the relative index of refraction. The absolute index is the ratio

$$n_{abs} = \frac{c}{v}, \qquad (4.28)$$

where c is the vacuum speed of light and v is the phase velocity of light in the medium. The relative index is the ratio of the speed of light in air to that in the material,

$$n_{rel} = \frac{v_{air}}{v_{medium}} = \frac{n_{abs}}{n_{air}}, \qquad (4.29)$$

where n_{air} is the index of air, which is about 1.0003 at standard temperature and pressure (Jones[123]).

There is a possibility for confusion since the symbol n is used both for relative and absolute refractive indices. If a ratio of indices is required, then both can be either absolute or relative. A measurement relative to air may be more appropriate for a component that will be working in air. In general, the 0.03% difference between absolute and relative index is more important in interferometry and less important in geometrical optics. For a glass plate 1 cm thick with a nominal index of 1.5, the difference in path length is 3 waves at $\lambda = 0.5$ μm. For a simple lens of the same material, whose nominal focal length is 100 mm, the difference between the air and vacuum focal lengths is 60 μm.

The indices of transparent solids are measured most accurately with a goniometer identical in form to a prism spectrometer, such as the high-precision instrument shown in Fig. 54. The sample material is made into a

FIG. 54. A precision prism goniometer. (Courtesy of Gaertner Scientific Corporation, Chicago, Ill.)

prism and mounted on the refractometer. At the wavelength of interest, a measurement is made of the minimum angle of deviation θ, which occurs for a beam that passes through the prism symmetrically (Fig. 55). The prism vertex angle α is measured on the same apparatus by autocollimation on

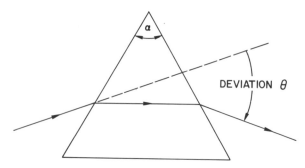

FIG. 55. The ray path through a prism at minimum deviation, which is symmetrical with respect to the prism apex.

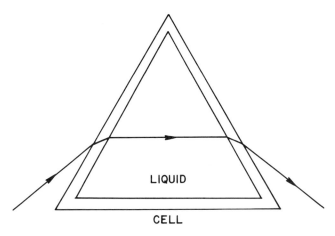

FIG. 56. Prismatic cell for liquids. If the cell walls are planar, the deviation depends only on the "liquid prism."

the prism faces, and the index is given by

$$n = \frac{\sin\left(\frac{\alpha+\theta}{2}\right)}{\sin\left(\frac{\alpha}{2}\right)}. \qquad (4.30)$$

The index can be measured to the sixth decimal place with extreme care, as discussed by Tilton.[124,125,126,127,128]. This requires that the goniometer degree wheels be accurate to a fraction of an arc second and that the prism be sufficiently large, well made, and precisely aligned. Temperature, humidity, and barometric pressure must be taken into account.

The index of a liquid can be measured in this way by filling a prismatic cell (Fig. 56). The cell walls must be plane and parallel, but their index is unimportant.

Another goniometric method (Fig. 57) requires only autocollimation. The index is found from measurements of the prism vertex angle α and of the angle θ between the normal to a near surface and the beam that is perpendicular, inside the prism, to the other surface.

The *Hilger-Chance refractometer*, or *vee-block refractometer* (Fig. 58), is used for more routine measurements, and accuracies of 0.00002 are possible. This is a *differential refractometer*, in which the output angle depends on both the index of the sample n_S and that of a component of the refractometer n_R. The sample index is found from the deviation of the beam passing through the instrument. Sample preparation and alignment are simpler than

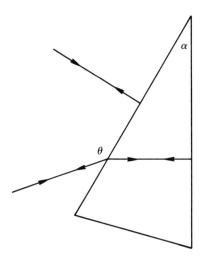

FIG. 57. The prism autocollimation method for index measurement. The index is found from the vertex angle and the angles made by the beams reflecting from the front and rear surfaces.

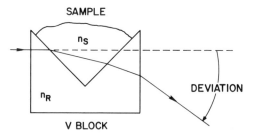

FIG. 58. Vee-block refractometer.

with prism spectrometers. Two faces of the sample are prepared at a 90° angle. A contact liquid between the vee and the sample increases the transmission across the interface and reduces the surface quality requirement of the sample. The index of a liquid can also be measured with this instrument.

The refractive index can be found from the critical angle associated with the planar interface between two materials of different indices, if one index is known. Figure 59 shows a block of material with a known high index, into which light from a sample of lower index is refracted if it is transmissive, or from which it is reflected if it is absorbing. To measure the cutoff angle, there must be a component of the illumination tangent to the interface. For

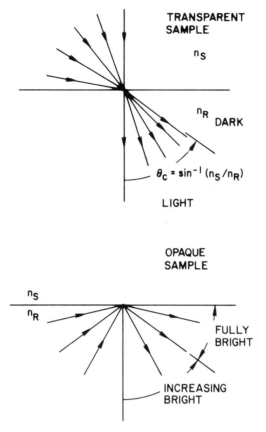

FIG. 59. Critical-angle methods of index measurement.

absorbing objects, illumination over a wide range of angles is necessary within the known substance. (The index of absorbing materials can be measured ellipsometrically.) *Critical-angle refractometers* are common in routine laboratory measurements of small amounts of liquids. To measure the index of a solid, only a single surface must be prepared, and a contact fluid with an index greater than that of the sample is used. Only the surface index of solids is measured, which may differ from the bulk index.

The *Abbe refractometer*[129,130,131] (Fig. 60) is a widely used critical-angle instrument. A liquid sample is held between two prisms, one of whose input faces is ground so that light is scattered into the liquid at all angles. If white light is used, the critical-angle edge is chromatic. Counter-rotating Amici prisms can achromatize the edge, at the same time measuring the sample's dispersion.[132] The accuracy of the Abbe refractometer can be as high as

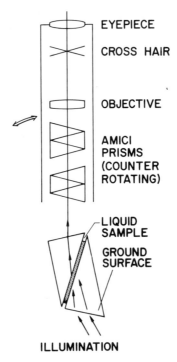

FIG. 60. The Abbe refractometer, the most common critical-angle instrument.

0.0001. Some instruments have a set of prisms that can be changed for increased range. The indices of many liquids depend strongly on temperature, so Abbe refractometers usually have tubes for a temperature-regulating fluid.

The *dipping* or *immersion refractometer* has a prism, like that of the Abbe refractometer, that is rigidly fixed to the telescope. It is immersed into the liquid, and the index is read from a reticle. This instrument has less range than the Abbe refractometer and requires more liquid.

Another critical-angle instrument is the *Pulfrich refractometer* (Fig. 61).[129,131,133] Its accuracy can be as great as 0.00002, better than that of the Abbe refractometer.

The prism goniometer can be used to make critical-angle measurements, as shown in Fig. 62. The prism base is ground so that the incident light is scattered at all angles. The index is found from the prism vertex angle and from the exterior angle at which the intensity of the light leaving the prism is maximum.

The index of an irregular solid sample can be found by immersing it in a liquid whose index is controlled by changing its temperature. The sample

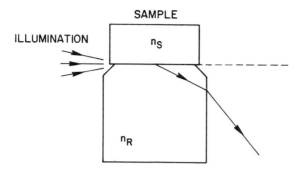

FIG 61. The Pulfrich refractometer, a critical-angle instrument.

vanishes when the two indices match. (A schlieren instrument can be used to judge its disappearance.) The index of a powder can be found in the same way, with there being no scattering when the particle and liquid indices are identical.

The index of individual particles can be determined microscopically by observation of the *Becke line*.[58] If a transparent particle is immersed in a liquid of a different index, then with defocus a dark line appears on one side of the interface and a light line on the other, depending upon the direction of defocus and upon which index is greater. The Nomarksi interference microscope can also be used to match the index of particles with that of their surrounding liquid,[134] as can the phase contrast microscope.

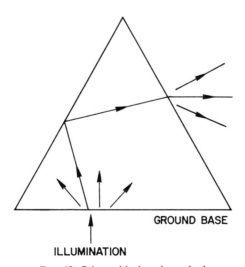

FIG. 62. Prism critical-angle method.

When a plane parallel plate is introduced in an optical system perpendicular to its axis, it shifts the image axially by an amount that depends on its thickness and index. Thus the index of a thin film of known thickness can be determined. This is called the *Duc de Chalnes method*.[58] Its accuracy is not great, because of the imprecision in locating a plane of best focus.

The index of very small quantities of a liquid can be found using the *Jelley refractometer*.[135] The sample is placed in a glass vee, through which the image of a slit source is viewed against a scale which is calibrated to give the index from the displacement of the image.

The refractive index of gases is usually measured interferometrically. Unlike liquids and solids, it is possible to begin with an evacuated chamber and add gas until a desired pressure is reached. As this is done, some change, usually a fringe movement, can be observed, giving the index. In principle, any two-beam interferometer can be used for this purpose, as can the Fabry–Perot etalon. (Candler[136] gives details.)

The first instrument of this sort was the *Rayleigh interferometer*[137] (Fig. 23, Chapter 5). Here light from a small aperture is divided into two beams that pass through identical chambers and are recombined to form a two-slit interference pattern. As the contents of one of the chambers is altered, its optical path length changes, and the fringe pattern shifts. Identical fringe patterns can be set into coincidence more accurately than the displacement of a single pattern can be measured. To make use of this principle, an additional interference pattern is produced with a parallel instrument. This reference pattern is held in registration with the changing pattern by tilting a calibrated compensation plate to vary an optical path length. The Rayleigh interferometer measures the group velocity, since sample dispersion shifts the position of the central fringe.

The Fourier transform spectrometer, discussed below, can be used for refractometry by placing a sample in one arm. An analysis of the interferogram gives the index.

An index-versus-wavelength curve can sometimes be produced directly. R. W. Wood[138] did so with sodium vapor by producing essentially a gas prism. A vertical gradient of vapor concentration was established by evaporating sodium at the base of a chamber and condensing it at the top. The abscissa of the dispersion curve was produced by crossing this arrangement with a prism that dispersed horizontally.

4.11. Spectroscopic Instruments

Spectroscopic instruments separate or analyze light according to wavelength λ or wave number $\sigma = 1/\lambda$. All spectrometers have a fundamental component that we call the *spectral unit*. In and near the visible region of the spectrum, this unit is usually a prism, a grating, a Fabry–Perot etalon, or a version of the Michelson interferometer. The instruments employing each of these are treated individually in the following sections. Since prism and grating instruments are very similar, their common features are discussed in the section on dispersing spectrometers. Discussed here are the aspects common to all spectrometers, a number of which are listed in Table II. All four spectrometer types are discussed in the books of James and Sternberg,[139] Bousquet,[140] Thorne,[141] Tarsov,[142] and Williams.[143] See also the article of Girard and Jacquinot.[144] Two older general references are Baly[145] and Clark.[146] Laser spectroscopy is not considered here.

The terminology of spectroscopic instruments as used here is as follows. *Spectroscopy* refers to the entire field, experimental and theoretical. *Spectroscope* is used both as a general term, and, as we use it here, to denote an instrument in which the spectrum is observed visually. In a *spectrograph* a portion of the spectrum is captured simultaneously on film. A more recent relative is the *multichannel spectrometer*, in which a fixed detector array simultaneously measures the power in a number of spectral intervals. A *scanning spectrometer* employs some changing element to produce a record, containing information from which the spectrum can be calculated. The record may be either nominally identical to the spectrum, or very different, as in the case of a Fourier transform instrument. A *monochromator* extracts a single spectral band from its input. A scanning monochromator followed by a detector is one type of scanning spectrometer. A *spectrophotometer* or *absorption spectrometer* measures the ratio of transmittance or reflectance in two channels. A *spectroradiometer* measures power as a function of wavelength. The term *spectrometer* is sometimes used to mean a spectral instrument other than a spectroscope or spectrograph; it is also used, as it is here, as a generic name for any spectral instrument.

Spectrometers may also be grouped by wavelength domain.[143] Another distinction is that made between emission and absorption work. Emission lines are bright against a dark background, while absorption lines are depressions in a continuum (Fig. 63), so throughput and signal-to-noise considerations are different.

Spectrometers can be thought of in a unified way as devices that divide and recombine light in such a way that optical path differences (OPD) are introduced.[147,148] The *order of interference* is the number of wavelengths of path difference, i.e., OPD/λ.

Spectral units are sometimes classified as either interferometric (Michelson and Fabry–Perot), as opposed to noninterferometric or dispersing (prism and grating). This distinction can be misleading, since all can be explained by interference and diffraction theory. However, there are

TABLE II. Spectroscopic Element Characteristics*

	Prism	Grating	Fabry-Perot	Michelson
number of beams	1	N	$\mathcal{F}$ (effective number)	2
means of division	none	wave front (breadth)	amplitude (length)	amplitude (length)
typical orders of interference	—	$\pm 1, 2, \ldots, 100s, 1000s$	1000s	1000s
OPD	$(d_1 - d_2)\delta n$	$MN\lambda$	$\sim 2ns\mathcal{F}$	X
resolving power, $\mathcal{R}$	$(d_1 - d_2)\, dn/d\lambda$	MN	$\mathcal{F}M_0$	$2X\sigma$
luminosity × resolving power, $\mathcal{L} \times \mathcal{R}$	—	—	$2\pi A$	$2\pi A$
free spectral range	—	$M\lambda_1$	λ/M	—
number of resolution elements per free spectral range	—	N	$\mathcal{F}$	—
instrument function	sinc^2	many sinc^2	Airy function	$\mathrm{sinc}(2\sigma X)$
input aperture	slit	slit	circle	circle
symmetry	bilateral	bilateral	circular	circular
scanning method	rotation	rotation	separation, pressure change	separation
action on plane wave by spectral unit	bends	bends, modulates, multiplies	modulates by angle, according to plate or mirror separation	
theoretical limitation on resolution	prism size and angle, material dispersion	number of lines	plate separation	length of travel
practical limitation on resolution	material homogeneity	ruling accuracy, flatness	plate flatness, coating	accuracy of motion and sampling

* See text for discussion and definition of symbols. Inapplicable quantities are left blank.

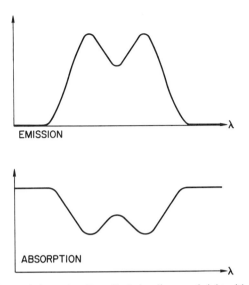

FIG. 63. Emission and absorption lines. Emission lines are bright with a dark background, while absorption lines are depressions in a bright background.

many ways in which the two pairs are similar. The Michelson and Fabry-Perot use "division of amplitude," in which an incident plane wave is divided by partial reflection in the same way everywhere across the wave front, giving two or more plane waves traveling in the same direction. Grating instruments use "division of wave front," in which the incoming wave front is acted on differently at different points.

Another classification of spectral units is by the number of divisions the incoming light undergoes, namely the number of "beams." The prism has a single beam. For a grating, the number is that of the lines. The etalon, which divides the light into a series of diminishing contributions, is assigned an effective number of beams by analogy to the grating. The Michelson interferometer is a two-beam device.

The overall form of most spectrometers is that of an imaging system in whose aperture stop is the spectral unit (Fig. 64). This arrangement forms a chromatically altered image of the entrance aperture or field stop. The field of view is defined by a field stop in the front focal plane of the collimating lens. The focusing lens, also called a camera or telescope, completes the imaging of the field stop. A concave grating performs both the spectral and imaging functions. In a spectroscope, the image is viewed with an eyepiece. In a spectrograph, the image is recorded by a photographic plate. Monochromators have an aperture in the image plane, and in scanning

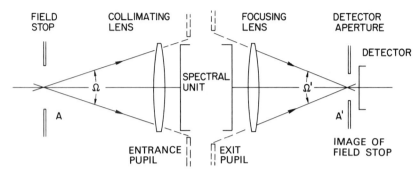

FIG. 64. General spectrometer layout. The field stop is imaged on the detector aperture by an optical system with the spectral unit in its stop.

spectrometers the aperture is followed by a detector. If the spectral unit is not the system aperture stop, then the unit is not fully utilized, since it is not filled by light from every point within the field stop.

A spectrometer can be thought of in another way. Spectral units—at least those that are entirely planar—operate on plane waves. Prisms change the directions of these waves according to wavelength; gratings multiply and disperse them; etalons and Michelson interferometers change the irradiance of the waves according to their angle and wavelength. The rest of the spectrometer is comprised of two parts. The collimating optics creates a plane wave from the light emanating from each point in the entrance aperture. The collimator is always required in a dispersion spectrometer, since an image of the entrance slit is formed, but it may not be necessary with an etalon or interferometer, since the light from the source can always be thought of as a superposition of plane waves. The focusing lens brings each plane wave leaving the spectral unit to a different point. This element is always necessary, since the output is ultimately analyzed spatially. Again, concave gratings combine all three roles.

Spectrometer performance is described by a number of properties, some of which are opposing, and not all of which apply to all instruments. The principal characteristics are power throughput, photographic speed, temporal speed, field of view, resolution, resolving power, throughput-resolving power product, spectral region, free spectral range, noise, signal-to-noise ratio, and information capacity. There are also less quantifiable criteria, such as simplicity and reliability. Most instruments have variable parameters and can be operated over a range of performance characteristics. There is no all-round best spectrometer; each measurement problem has its own optimum solution.

The *power throughput* or power-transmitting capacity of a spectrometer is the ratio of output flux to input radiance. It depends upon the geometry of the spectrometer and upon the transmissivities and/or reflectivities of its elements. The geometrical factor is

$$A\Omega = A'\Omega', \qquad (4.31)$$

where A is the area of the field stop and Ω is the solid angle subtended at the field stop by the entrance pupil. A' and Ω' are the corresponding quantities on the output side. This equality is an expression of the optical invariant [Eq. (2.27)] and the conservation of brightness. The geometrical factor can also be written as a product of the area of the spectral unit, and the solid angle subtended there by the field stop. If τ is the system transmittance factor that accounts for absorption and scattering, the throughput factor is

$$\mathscr{L} = A\Omega\tau = A'\Omega'\tau. \qquad (4.32)$$

This quantity is also called the *luminosity, etendue, brightness, collection power, light gathering power*, etc. These terms are sometimes applied only to the geometrical portion of this expression. If the incoming radiation fills the field stop and the aperture stop uniformly, then the power that can be transmitted by the spectrometer is

$$P = \mathscr{L}L, \qquad (4.33)$$

where L is the radiance (power/area/solid angle) of the incoming light. With photoelectric detectors, the important quantity is the total power incident on the detector. Film can be thought of in the same way, with each grain a detector; but exposure requirements are usually thought of in terms of energy per area, which is increased when the magnitude of the magnification of the entrance aperture is decreased. Equivalently, this occurs when the focal length of the focusing lens is reduced, giving a larger value for Ω', that is, a "faster" system.

No spectrometer can determine the spectrum with absolute fidelity. Each has a characteristic *instrument function* or *instrument profile*, its output for a monochromatic input. If this function is the same for all wavelengths, then, aside from noise, the actual output is the convolution of the instrument function and the nominal output. Typically, the instrument function is sufficiently unchanging over a broad range of wavelengths to be used in this way. The instrument function may be normalized to unit height or unit area, or the inefficiencies contributing to τ may be included. The instrument function can be measured with a narrow band input.

Over a spectral region where the instrument function does not change, the spectrometer acts as a linear system. So, if the input power varies sinusoidally with wavelength or wave number (a so-called channeled spectrum), the output is also sinusoidal, but with less modulation. This ratio of the output to input modulations is given by the *spectrometer transfer function*,[149,150,151] the normalized Fourier transform of the instrument function. This quantity is analogous to the optical transfer function of noncoherent imaging systems.

The *spectral resolution*, $\delta\lambda$ or $\delta\sigma$, is the difference of wavelength or wave number required of an isolated pair of narrow emission lines of equal power in order that they can be recognized as separate. The resolution depends on the width of the instrument function, which decreases as the OPD increases. Several resolution criteria are used, including the full width of the instrument function at half maximum, the separation required to give a dip in the two-wavelength output, and the position of the first zero of the instrument function. The numerical values for resolution associated with these criteria are very close. (These resolution criteria are analogous to those applied to imaging systems for spatial resolution.) The *spectral purity* or *filtrage* is the monochromator analogue to resolving power. One definition of this quantity is the ratio of the integrated power in the desired spectral region to that from all other regions.

The *resolving power* is defined as

$$\mathscr{R} = \frac{\lambda}{\delta\lambda} = \frac{\sigma}{\delta\sigma}. \qquad (4.34)$$

A rough limit on resolving power can be found from the quantum mechanical time-energy uncertainty relationship,

$$\delta E \, \delta t \geq h, \qquad (4.35)$$

where δE is the energy uncertainty, δt is the time uncertainty, and h is Planck's constant. The spectral unit reduces the energy uncertainty by increasing the uncertainty in the time that a photon will leave the system, accomplishing this by dividing and recombining the light so as to introduce optical path differences. The time uncertainty is related to the maximum path difference by

$$\delta t = \frac{\text{OPD}}{c}, \qquad (4.36)$$

where c is the speed of light. Since $E = h\nu$, where $\nu = \sigma c$,

$$\delta E = hc\delta\sigma. \qquad (4.37)$$

The uncertainty principle gives as a lower limit

$$\text{OPD}\, \delta\sigma \sim 1, \qquad (4.38)$$

and

$$\mathcal{R} \sim \frac{\text{OPD}}{\lambda} = \text{OPD}\ \sigma. \qquad (4.39)$$

The OPD depends on some dimension of the spectrometer, as shown in Table II.

A distinction must be made between the resolving power and instrument function of a spectrometer and that of a spectral element taken alone, i.e., its *intrinsic resolving power*. The resolving power of the spectral unit is always greater than that of the spectrometer in which it is used.

Increased resolving power is usually associated with decreased throughput, and many instruments have a characteristic luminosity-resolving power product, $\mathcal{L} \times \mathcal{R}$ (Fig. 65). A number of instruments do not have such a product, for instance, the field-widened Fourier transform spectrometer, the spherical Fabry-Perot etalon, and the spectrograph.

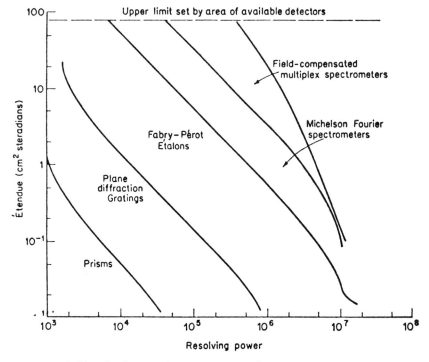

FIG. 65. Limiting $\mathcal{L} \times \mathcal{R}$ curves for common types of spectrometers, as of 1969. (Courtesy of J. F. James and R. S. Sternberg, *The Design of Optical Spectrometers* (Chapman and Hall, London, 1969).)

Spectral units are usually circular or rectangular. Dispersion instruments usually have input slits and bilateral symmetry, while Fabry-Perot and Michelson spectrometers have axial symmetry and use circular field stops. Thus for the same resolution, the latter have greater throughput, an increase called the *Jacquinot advantage*.[152]

Each spectrometer has wavelength range limitations determined by its geometry and the materials of which it is composed. There are always limits at short wavelengths imposed by mechanical tolerances, which are small fractions of a wavelength. As is required at the wavelengths of interest, the materials must be transmissive, reflective, dispersive, or homogeneous. The spectral response of detectors is another limiting factor. For example, the spectrograph cannot operate at wavelengths beyond which film is responsive, i.e., about 1.2 μm. Some instruments use interchangeable components to extend their ranges.

If there are numerous major peaks in the instrument function, as with gratings and etalons, then only a finite wavelength range can be used without overlap. This is called the *free spectral range*. An instrument with a small free spectral range may require auxiliary band-limiting apparatus or a spectrometer in series to separate the overlapping regions. Other important spectrometer characteristics are the number of resolvable spectral elements in the total range and the number of resolvable elements in the free spectral range.

Noise afflicts spectrometers, which are sometimes detector noise-limited and sometimes photon noise-limited. Scanning spectrometers are susceptible to a particular type of error due to source fluctuations during the scan. Fourier spectrometers have unique forms of noise, since the noisy signal is Fourier-transformed to give the spectrum. A lower signal-to-noise ratio is required to resolve emission lines with no continuum than for absorption lines. The signal-to-noise referred to here is measured with respect to the signal maximum.

A performance measure introduced by Connes[153] is

$$Q = (\text{number of spectral elements}) \times (\text{signal-to-noise ratio}), \quad (4.40)$$

which equals the ratio of the total energy in the spectrum to that in the smallest detectable line, one for which the signal-to-noise ratio is unity. This quantity is also related to the *information capacity* of the spectrometer, the product of the number of resolved elements and the number of distinguishable irradiance levels.

The terms *multiplex*, *multiplex advantage*, and *Fellgett advantage*[154] are applied to instruments in which the detector simultaneously measures light in more than one wavelength region. The advantage is gained if the measurement is limited by signal-independent detector noise. For signal-dependent noise there may be a disadvantage. The degree of the multiplex advantage is proportional to the square root of the number of resolvable spectral elements simultaneously measured, called the *multiplex factor*; so with narrow-band sources there is little gain. The multiplex advantage can be used either to improve the signal-to-noise ratio for measurements made in a given time or to reduce the measurement time for a given signal-to-noise ratio.

Scattered light degrades spectrometer performance.[155] Since absorption lines are dips in a bright background, scattered light is generally more troublesome in absorption work than in emission spectroscopy, with the exception of Raman spectroscopy, where the lines are weak. An avoidable source of scattering is overfilling the spectral unit. Effects of scattering can be reduced by using spectrometers in series, so that only a fraction of the incoming light enters the second spectrometer.

Several types of series arrangements are used in spectroscopy. Within a spectrometer, multiple spectral units are used, and single units are used in multiple passes. Complete spectrometers are used in series in order to increase resolution, to reduce scattered light, and to remove the light beyond the free spectral range of one of the instruments. When two dispersion instruments are used in series, their dispersions add. Dispersion systems or elements in series can be crossed to disperse in orthogonal directions, thus separating otherwise overlapping spectral regions. The luminosity of a series of spectrometers is geometrically limited by that with the lowest etendue. In order to conserve energy, reduce scattering, and obtain uniformity across the output field, a field lens may be required at the common field apertures of spectrometers in series to image the spectral unit of each spectrometer into that of the other.

The method of spectrometer illumination depends upon the shape of the source, its brightness, and the angular distribution of its radiation. For a given source, throughput is maximized if both the field stop and the aperture stop of the spectrometer are filled, which requires that the optical invariant of the source and illumination system equals or exceeds that of the spectrometer. If both spectrometer stops cannot be filled, a trade-off can be made between area and angle. For instance, uniform slit filling is preferred if line strengths are to be found by measuring photographs of the spectrum. However, resolution can be reduced if a grating or prism is not filled. The basic illumination schemes are identical to those for projectors discussed above; Koehler, critical, and diffuse illuminations are used (although spectroscopists do not use these names). A source that is sufficiently large and that radiates over a wide enough range of angles to fill both stops requires no illumination optics; and no optics can increase the output power, since an optical system cannot increase brightness. Smaller sources are imaged

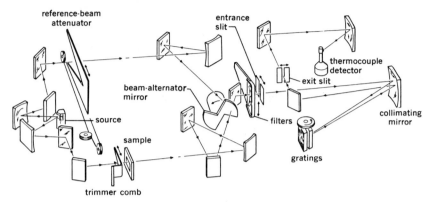

FIG. 66. A spectrophotometer. The reference-beam and sample-beam powers are equalized by physically truncating the reference beam with a triangular attenuator. (Courtesy of Beckman Instruments, Fullerton, Ca.)

onto the field stop in critical illumination, for which the condenser should be sufficiently fast to fill the spectrometer aperture, and a lens at the field stop may be required to image the condenser onto the spectrometer aperture so that no radiation is lost. Koehler illumination can be used with a nonuniform source to evenly fill the field stop. Light from sources whose shapes are very different from the spectrometer apertures is sometimes spatially rearranged.[156]

Spectrophotometers are discussed by Bauman,[157] Driscoll,[158] and in a number of analytical chemistry books. Most commercial absorption spectrometers are self-contained instruments, as the example in Fig. 66 shows. The source has a continuous spectrum, and a monochromator either follows or precedes the sample. The light is alternately passed through the sample and the reference, and the ratio is measured. If the reference arm is empty, the absolute transmittance of the sample is obtained. To avoid depending on detector linearity, the ratio is often found with a nulling system using slits of variable width, triangular stops, or variable-transmissivity neutral density filters.

4.11.1. Dispersion Spectrometers

In this section we treat the features common to dispersion spectrometers, whose spectral elements are prisms and gratings. These instruments are discussed by Strong,[159] Sawyer,[160] Harrison et al.,[161] Meltzer,[162] Stroke,[163] Girard,[164] and Girard and Jazquinot.[144]

A prism or grating is characterized by its size, transmissivity or reflectivity, resolution, resolving power, and *angular dispersion*, $d\theta'/d\lambda$ or $d\theta'/d\sigma$, where

SPECTROSCOPIC INSTRUMENTS   201

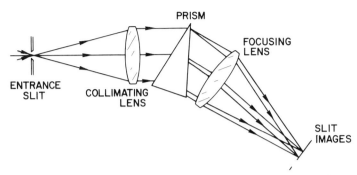

FIG. 67. The Prism spectrometer. Spatially separated slit images are formed in the various wavelengths of the incoming light.

θ' is the outgoing angle. Large dispersion is necessary for high resolution, but does not guarantee it. For a given resolution, an instrument with a high-dispersion element can be more crudely made. In general, angular dispersion varies with wavelength. Wavelength-independent dispersion is called *normal dispersion*.

The basic form of dispersion spectrometers is that of the prism instrument shown in Fig. 67. Light from points in the entrance slit is collimated by a mirror or an achromatic lens and enters the prism or grating, from which the various spectral components are sent in different directions. The outgoing beams are focused by a second lens or mirror, giving spatially separated slit images in the various wavelengths. In *multiple-pass* spectrometers, the light is acted on by the disperser more than once. The common *Littrow arrangement* (Fig. 68) is folded so that a single lens both collimates the light and focuses it near the entrance slit.

Ideally, the imaging should be stigmatic, so that each point in the entrance slit is imaged as a point. In some spectrometers, particularly those with nonplanar dispersing elements, e.g., concave gratings, the imaging is strongly astigmatic. In this case, the output surface is taken to be the astigmatic focal surface on which the focal lines lie in the direction of the geometrical slit image.

Most dispersion monochromators have an output slit, that is fixed in position, from which a selected narrow-band output is obtained by rotating the disperser. In scanning dispersion spectrometers, a detector follows the output slit.

Two spectral lines are resolved if the slit images in the two wavelengths are sufficiently separated. The geometrical separation of the slit images depends on the angular dispersion of the spectral component, the focal length of the focusing lens, and the width of the entrance slit. The actual

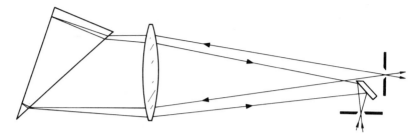

FIG. 68. The Littrow arrangement with a prism used in double pass. Here one side of the prism is reflectively coated. Alternately, a separate mirror behind the prism can be used.

image depends also on diffraction effects, whose dimensions depend on the numerical aperture of the image-forming cones.

A purely geometrical analysis suffices if the slit width is large compared to the dimensions of the diffraction effects. Slit images at wavelengths that differ by $\delta\lambda$ or $\delta\sigma$ have a center-to-center separation,

$$\delta x' = f' \, \delta\theta' = f' \frac{d\theta'}{d\lambda} \delta\lambda = f' \frac{d\theta'}{d\sigma} \delta\sigma, \qquad (4.41)$$

where f' is the focal length of the focusing lens. The slit images just touch if this distance equals their width w', giving the condition for resolution,

$$w' = f' \frac{d\theta'}{d\lambda} \delta\lambda = f' \frac{d\theta'}{d\sigma} \delta\sigma. \qquad (4.42)$$

Thus, the just-resolved spectral lines are separated by

$$\delta\lambda = \frac{w'}{f'} \frac{d\lambda}{d\theta'} \quad \text{or} \quad \delta\sigma = \frac{w'}{f'} \frac{d\sigma}{d\theta'}, \qquad (4.43)$$

and the resolving power is

$$\mathcal{R} = \frac{f'}{w'} \lambda \frac{d\theta'}{d\lambda} = \frac{f'}{w'} \sigma \frac{d\theta'}{d\sigma}. \qquad (4.44)$$

These equations can also be written in terms of the input slit width and the collimator focal length, since $f/w = f'/w'$. Instead of angular dispersion, spectrometers are sometimes described by the *reciprocal linear dispersion*, the rate of change of wavelength or wave number with position in the output plane, $d\lambda/dx' = 1/f'(d\theta'/d\lambda)$ or $d\sigma/dx' = 1/f'(d\theta'/d\sigma)$. This quantity is called the *plate factor*.

To obtain the highest resolution, the entrance slit is so narrowed that its image is predominantly determined by diffraction. The image width then

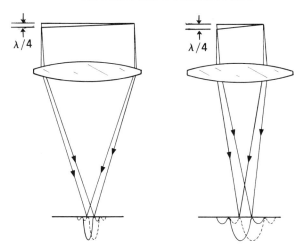

FIG. 69. The Rayleigh resolution criterion for spectral resolution, as applied to cases of different beam width.

depends on the numerical aperture of the focusing lens. According to the Rayleigh criterion,[165] slit images are just resolved if the maximum of each is located at the first zero of the other. This occurs if the two wavefronts are tilted relatively by one wavelength as they leave the dispersing element. This criterion is independent of beam width, since for tilts that are identical in this way, narrow beams have a greater angular separation than wide beams, but their diffraction patterns are correspondingly broader (Fig. 69). If b' is the width of the beam leaving the dispersing element, then for wavelengths around λ that differ by $\delta\lambda$, beams with a one-wave tilt differ in direction by

$$\delta\theta' = \frac{\lambda}{b'} = \frac{d\theta'}{d\lambda}\delta\lambda = \frac{d\theta'}{d\sigma}\delta\sigma. \quad (4.45)$$

This gives the resolving power

$$\mathcal{R} = b'\frac{d\theta'}{d\lambda} = b'\sigma^2\frac{d\theta'}{d\sigma} = \frac{b'}{\lambda}\lambda\frac{d\theta'}{d\lambda} = \frac{b'}{\lambda}\sigma\frac{d\theta'}{d\sigma}. \quad (4.46)$$

This is the intrinsic resolving power of a dispersing spectral element. The resolving power of a spectrometer using such an element can only be less. The angular dispersion term $d\theta'/d\lambda$ is related to the nature of the dispersion element, i.e., the period of a grating or the material and shape of a prism. This term determines the separation of the centers of the slit images. The term b' has to do with the size of the dispersion element and is related to the width of the slit image. Both the width of the diffraction pattern and the separation of the centers of slit images at different wavelengths are proportional to f', so that f' does not appear in the expression for $\mathcal{R}$. Comparing the dispersion spectrometer with an imaging system, b'/f' is like the NA, and $f'\, d\theta'/d\lambda$ is like transverse magnification. Thus a spectrometer with high dispersion and low resolution is analogous to an imaging system with empty magnification; the centers of the slit images at the various wavelengths are well separated, but the images are proportionally wider.

As the slit width is increased, starting from zero, resolution does not fall appreciably until the width of the geometrical image reaches the size of the central maximum of the linespread function, i.e., $w' = f'\lambda/b'$. This can be written as $w'/\lambda = f'/b'$, which means that the width of the geometrical slit image measured in wavelengths equals the F/number of the imaging-forming bundle. A discussion on the effects of slit width is given by Jacquinot and Dufour[166] and Stroke.[163] The coherence of the light entering the slit affects its image somewhat and has a small effect on the resolution.[167]

A trade-off between power and resolution determines the slit width. In the geometrical regime, Eq. (4.44) gives

$$\mathcal{L} \times \mathcal{R} = \tau \frac{h'S}{f'} \lambda \frac{d\theta'}{d\lambda} = \tau \frac{h'S}{f'} \sigma \frac{d\theta'}{d\sigma}, \qquad (4.47)$$

where h' is the slit image height, S is the illuminated area of the output side of the dispersing element, and τ is the transmittance factor accounting for absorption and scattering. In the diffraction regime, as the slit is narrowed, the resolution approaches a fixed value determined by diffraction spreading, but the throughput is diminished to zero, so $\mathcal{L} \times \mathcal{R}$ decreases. Maximum $\mathcal{L} \times \mathcal{R}$ occurs when the width of the geometrical image of the slit approximately equals the width of the diffraction spread function. In a spectrograph, film grain size determines the lower limit on the width of the slit image. The important quantity here is the power/area in the slit image, rather than the total power. As a result, for a spectrograph $\mathcal{L} \times \mathcal{R}^2 = \tau \mathcal{R}_0^2 (\lambda/g)^2$, where $\mathcal{R}_0$ is the intrinsic resolving power of the spectral unit and g is the film grain size. Bousquet[140] gives a more thorough discussion.

The instrument function in dispersion spectrometers is determined primarily by the slit image. For a spectrograph, it is the convolution of the slit image with the spread function of the film. In a scanning spectrometer, the instrument function is the convolution of the slit image and the detector aperture. In the geometrical regime, rectangular source and detector apertures give a trapezoidal or triangular instrument function (Fig. 70). The full width of the instrument function is the sum of the detector width and input slit image widths. For a given sum, throughput is maximized if the two widths are equal, in which case the instrument function is triangular. The

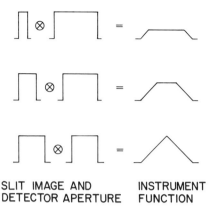

FIG. 70. The geometric monochromator instrument function, the convolution of the geometric slit image and the detector aperture. Three cases are shown for which the sum of the two slit widths is identical. Equal slits give greatest throughput.

wavelength distribution from a monochromator is a function of the same form. The bandwidth is the product of the plate factor and the width of the instrument function.

Aberrations play a fundamental role in spectrometer design. Some aberrations arise from the dispersion element, and others are due to the collimating and focusing optics (see Murty[168] for more discussion). The importance of aberrations varies with the technique being used.

There are a number of considerations relating to spectrometer slits. Since dispersion usually varies with wavelength, if the instrument function of a monochromator or scanning spectrometer is to be wavelength-independent, the slit width must be varied as the central wavelength is changed. Variable-width slits are preferably bilateral, opening symmetrically, so that their center positions do not change. Dispersion instruments usually give curved slit images, so the input slit, output slit, or both may be curved to compensate. There are a number of techniques involving spectrograph slits. Two or more identifiable spectra can be superimposed on a spectrograph plate for comparison or calibration by changing the slit height or by shifting the slit longitudinally between exposures. In order to measure the relative intensities of spectral lines with a spectrograph, a variation in transmittance with position along the length of the slit can be produced with a gradient neutral-density filter or with a spinning sector wheel that simulates such a filter in a time exposure. With a high-contrast plate, the length of the recorded image of a spectral line then depends on its intensity. This method requires uniform slit illumination and stigmatic imaging.

In simple dispersion spectrometers, power throughput at high resolution is limited by the need for narrow slits. Several methods have been devised to improve upon this situation by using enlarged input and output apertures of special design. Girard[164] uses identical large input and output apertures, whose correlation has a narrow central region and a broad base. The opaque portions of the output aperture are reflective. Separate detectors measure the total reflected and transmitted signals, which are identical except for a narrow peak, and the two signals are subtracted to give a sharp instrument function.

The method of Golay[169,170] employs an array of entrance slits, exit slits, or both. As the spectrum is scanned, a single detector measures the total power, which for each scan position is a linear combination of contributions from various wavelengths. The slits are arranged so that the combinations are linearly independent, and thus the power at each wavelength can be calculated from a number of points on the scan record. The dynamic range required of the detector is reduced if the slits are arranged so that, for a flat spectrum, the output does not change during the scan. In Hadamard transform spectroscopy,[171,172] the arrangement is also optimized for fast numeric inversion.

4.11.2. Prism Spectrometers

Dispersing prisms and prism spectrometers are discussed by Kingslake[173] and in most of the references given above.

The angular deviation and dispersion of a prism depends on its material, its apex angle, and the azimuth of the incoming beam. The general expressions for deviation and dispersion are cumbersome, but simple equations obtain for the symmetrical minimum deviation configuration, which is commonly used (Fig. 55). Equation (4.30) gives the relation between the minimum deviation angle, the prism apex angle α, and the prism index n. The dispersion at minimum deviation is

$$\frac{d\theta'}{d\lambda} = \frac{2\sin(\alpha/2)}{\sqrt{1-n^2\sin^2(\alpha/2)}}\frac{dn}{d\lambda} = 2\tan\left(\frac{\alpha+\theta}{2}\right)\frac{1}{n}\frac{dn}{d\lambda}. \quad (4.48)$$

The quantity $dn/d\lambda$ is the dispersion of the prism material. Since the index is not a linear function of wavelength or wave number, a prism spectrometer must be calibrated. If $\alpha = 60°$ and n lies between 1.4 and 1.6, then $d\theta'/d\lambda = n\,dn/d\lambda$ to within 4%.

For each index, there is an apex angle beyond which all the light that enters the first face it totally internally reflected at the second face,

$$\alpha_{\max} = 2\arcsin\left(\frac{1}{n}\right). \quad (4.49)$$

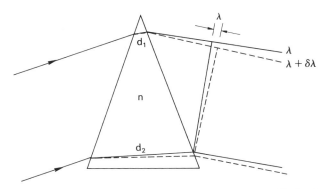

FIG. 71. Diagram for deriving prism resolving power. Two beams differing in wavelength by $\delta\lambda$ are relatively tilted by λ.

As α approaches α_{max}, the dispersion at minimum deviation becomes infinite, but the beam width approaches zero, so $\mathscr{L} \times \mathscr{R}$ approaches a finite limit.

The Rayleigh criterion for resolving power of a prism can be found from Fig. 71. Since the refractive index of air is nearly unity, two beams of nearly the same wavelength are tilted by one wave upon leaving the prism if the differences in their optical path lengths through the thick and thin portions of the prism differ by one wave. If d_1 is the lesser of the two lengths through the prism and d_2 is the greater one, this difference is

$$\text{OPD} = (d_2 - d_1)\delta n = (d_2 - d_1)\frac{dn}{d\lambda}\delta\lambda, \quad (4.50)$$

where δn is the difference in refractive index for the two wavelengths, which differ by $\delta\lambda$. Applying the Rayleigh criterion, the resolving power is obtained by setting OPD equal to λ, giving

$$\mathscr{R} = (d_2 - d_1)\frac{dn}{d\lambda} = (d_2 - d_1)\sigma^2 \frac{dn}{d\sigma}. \quad (4.51)$$

The index often varies with wavelength according to the Cauchy approximation $n(\lambda) = a + b/\lambda^2$. In this case, $\mathscr{R} \propto 1/\lambda^3$ and $\delta\lambda \propto \lambda^2$.

For several prisms in series or for multiple passes through a single prism, the net resolving power is a sum of such terms. The highest resolving power is obtained with a given prism if the beam is wide enough to fill it. For a filled prism working at minimum deviation, $d_2 - d_1$ equals the width of its base. Prism height, the length along the vertex, affects throughput, but not resolution.

In each wavelength range, resolving power is limited by the dispersion of suitable prism materials and by the size of optical-quality pieces that can be obtained. Light from the far ultraviolet to the far infrared can be analyzed by prisms of fluorite (CaF_2), quartz, glass, and alkali halides. Between 0.2 μm and 0.6 μm, the range of available dispersions is 1.6/μm to 0.04/μm. For example, Schott glass SF-6 has an index difference of 0.032 between the hydrogen F and C lines ($\lambda = 0.486$ μm, 0.656 μm) and an average dispersion in this region of 0.19/μm. At $\lambda = 0.5$ μm, such a prism with a 1 cm base has a resolving power of about 2000 and a resolution of about 2.5 Å. For an apex angle $\alpha = 60°$, its angular dispersion at minimum deviation is about 24°/μm.

A prism cannot be used at its maximum possible dispersion, since the conditions that give highest dispersion also result in the greatest flux loss. Material dispersion is greatest at those wavelengths were absorption is increased. In addition, for large prism apex angles, incidence angles are high, so Fresnel reflection losses are great. (In general, prisms act as partial polarizers because of Fresnel losses.)

A prism working in collimated light introduces no aberrations. However, points in the slit off the principal meridian of the collimating lens give rise to beams that pass through the prism at a compound angle. For these beams, the prism vertex angle is effectively increased, so they are deviated more than those in the principal meridian, and the slit image is curved in the direction of greater dispersion, toward the short-wavelength end of the spectrum. Compensation by curved slits is imperfect, since the curvature is wavelength-dependent. With noncollimated beams, prisms introduce severe aberrations, especially astigmatism.

The Littrow arrangement (Fig. 68), common in monochromators and scanning spectrometers, is more compact than an equivalent unfolded system and requires less prism material. The back of the prism may be reflectively coated or there may be a separate mirror. The optical activity of a crystalline quartz prism is cancelled in the Littrow mode. A disadvantage of the Littrow arrangement is that light scattered by the lens or prism can easily reach the exit slit.

In *constant-deviation spectrometers*, the direction of the outgoing beam at the wavelength of minimum deviation is fixed, along with the collimating and focusing lenses, and the wavelength is changed by rotating the prism. The *Pellin-Broca prism* (Fig. 72) can be thought of as an isosceles prism, folded to give a deviation angle of 90°. In the *Wadsworth* arrangement (Fig. 73), a mirror is rigidly attached to the prism, and the entire unit rotates. The deviation angle is $\theta = 180° - 2\phi$, where ϕ is the angle between the mirror surface and the bisector of the prism apex. This constant-deviation property is analogous to that of a pair of plane mirrors, since a prism at minimum

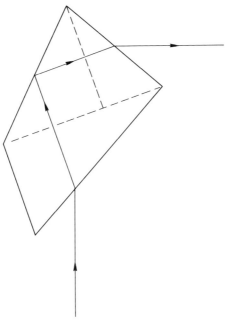

FIG. 72. The Pellin-Broca constant-deviation prism, equivalent to a 60° prism that has been "rearranged." Light at the wavelength that is minimally deviated is deviated by 90°. With the collimating and focusing optics fixed at a 90° angle, light at the minimally deviated wavelength is directed to the focusing optics by rotating the prism.

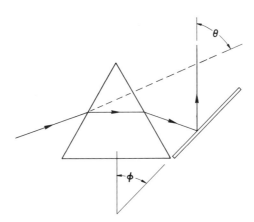

FIG. 73. The Wadsworth constant-deviation arrangement. A mirror is rigidly attached to a prism, and the pair rotate together. The deviation angle for light at the minimally deviated wavelength is $\theta = 180° - 2\phi$, where ϕ is the angle between the prism bisector and the mirror.

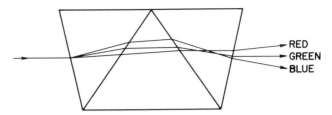

FIG. 74. The Amici prism, which does not deviate light at a central wavelength.

deviation is symmetric about the bisector of its vertex, just as a mirror is about its surface.

The *direct-viewing prism* consists of simple prisms arranged in opposition so that the spectrum is spread about an undeviated central wavelength. The Amici prism (Fig. 74) is one such arrangement. Such prisms are used chiefly for low-resolution visual devices, such as hand-held spectroscopes.

4.11.3. Grating Spectrometers

The basic theory of the grating is given in Chapter 6, and Stroke[174] gives a comprehensive treatment. The shorter articles by Loewen,[175,176] Richardson,[177] Kneubuhl[178] and Schroeder.[179] Davis,[180] Sawyer,[160] and Harrison et al.[161] discuss grating spectrometers. Beutler[181] and Welford[182] treat grating aberrations. Holographic gratings are discussed by Schmahl and Rudolph.[183]

For a plane grating with a plane wave input, there are output power maxima at angles for which the optical path difference between adjacent periods is an integral number of wavelengths (Fig. 16, Chapter 6). This integer is called the *order of interference*. For an incident plane wave with a direction vector in a plane perpendicular to the grating lines at an angle θ from the normal to the grating surface, the outgoing angle θ' in the Mth order is given by the *grating equation*

$$n' \sin \theta' = n \sin \theta + M \frac{\lambda}{p}. \quad (4.52)$$

Here p is the grating period, λ is the vacuum wavelength, and n and n' are the indices of the two spaces, which usually equal unity. Figure 75 shows for a transmission grating the outgoing directions for the positive integer orders for a normally incident beam containing the wavelengths $\lambda_1 = p/5$ and $\lambda_2 = p/4$. For order numbers such that $|\sin \theta'| > 1$, there is an evanescent wave, which does not leave the grating. The number of propagating orders is twice the integral part of p/λ, plus or minus one; and for $\lambda > 2p$, there can only be a zeroth order. If the incident direction vector is not in a plane normal to the grating lines, then the period is effectively decreased by the

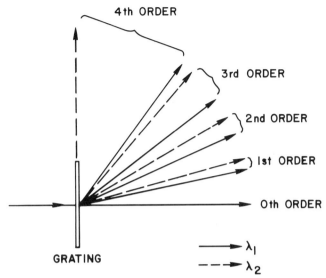

FIG. 75. Grating output angles for normal incidence with wavelengths $\lambda_1 = p/5$, $\lambda_2 = p/4$, where p is the grating period.

cosine of the angle this vector makes with the perpendicular to the lines, and the diffraction angle is reduced. Thus the slit image in a grating spectrometer is curved in a direction opposite to that of a prism spectrometer.

The angular dispersion, obtained by differentiating the grating equation, is

$$\frac{d\theta'}{d\lambda} = \frac{M}{pn'\cos\theta'} = \frac{n'\sin\theta' - n\sin\theta}{\lambda n'\cos\theta'} \quad \text{or} \quad \frac{d\theta'}{d\sigma} = \frac{n'\sin\theta' - n\sin\theta}{\sigma n'\cos\theta'}.$$
(4.53)

Note that the dispersion can be written without explicit dependence on the order number. The dispersion is almost constant for outgoing directions that are nearly normal to the grating. For a normal incident beam, the dispersion can be written

$$\frac{d\theta'}{d\lambda} = \frac{\tan\theta'}{\lambda} \quad \text{or} \quad \frac{d\theta'}{d\sigma} = \frac{\tan\theta'}{\sigma}.$$
(4.54)

The resolving power for a grating of width W is found from Eq. (4.46) and Eq. (4.53). Here $b' = W\cos\theta'$ is the width of the outgoing beam. For a grating with N grooves, $W = Np$. The resolving power is

$$\mathcal{R} = |M|N = \frac{W}{\lambda}|n'\sin\theta' - n\sin\theta| = \frac{\text{OPD}}{\lambda},$$
(4.55)

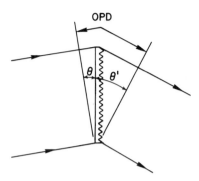

FIG. 76. Maximum optical path difference for a particular diffraction order. Its magnitude depends on the angles the order makes with the grating, the width of the grating, and the refractive indices.

where the OPD here is the difference between path lengths along the extremes of the incident and diffracted beams (Fig. 76). A coarse grating working in a higher order can have the same dispersion and resolution as that of a finer grating of the same size in a lower order. The angular dispersion becomes infinite as θ' approaches 90°, but the beam width approaches zero, so the resolving power is bounded. The maximum resolving power, obtained in the limit of 180° deviation, is

$$\mathcal{R}_{\max} = \frac{2W}{\lambda} = 2W\sigma. \qquad (4.56)$$

Thus the limiting wave number resolution is $\delta\sigma = 1/2W$. As with a prism, the height of a grating, the length along its grooves, affects throughput but not resolving power.

In the Littrow configuration (Fig. 77), $\theta' = -\theta$, so the dispersion is $d\theta'/d\lambda = 2\tan\theta'/\lambda$, and the resolving power is $\mathcal{R} = 2W\sin\theta/\lambda$. A high resolving power is possible with a narrow beam.

Various wavelengths in different orders can have diffraction maxima at the same angle. The free spectral range is the portion of the spectrum in a given order not overlapping with that from other orders. Since dispersion increases with the magnitude of the order number, the first intrusion into a given order is from the light of the next highest order. Thus for order M, if λ_1 is the shortest wavelength in the free spectral range and λ_2 is the longest, then

$$|M|\lambda_2 = (|M|+1)\lambda_1. \qquad (4.57)$$

Therefore, if $n = n' = 1$, the ratio of the extreme wavelengths in the free

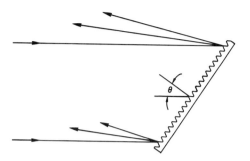

FIG. 77. The grating Littrow arrangement, in which the direction of the observed light is nearly opposite that of the incoming light.

spectral range is

$$\frac{\lambda_2}{\lambda_1} = \frac{|M|+1}{|M|}. \tag{4.58}$$

The difference between the extreme wavelengths is

$$\lambda_2 - \lambda_1 = \frac{\lambda_1 + \lambda_2}{1 + 2|M|} = \frac{\lambda_1}{|M|} = \frac{\lambda_1 N}{\mathcal{R}} = N\delta\lambda = \frac{\lambda}{p \cos \theta' \dfrac{d\theta'}{d\lambda}}. \tag{4.59}$$

The difference in wave numbers is

$$\sigma_1 - \sigma_2 = \frac{\sigma_2}{|M|} = \frac{\sigma_2 N}{\mathcal{R}} = N\delta\sigma. \tag{4.60}$$

Thus the free spectral range decreases with order number, dispersion, and resolving power. For a given resolution, the free spectral range is proportional to the number of grating lines.

Order sorting, the prevention of overlapping orders, is accomplished by limiting the spectral range with a filter or with a low-dispersion instrument in series. In scanning spectrometers and spectrographs, overlap may be effectively eliminated by the insensitivity to some wavelengths of the detector or photographic plate. Without limiting the spectrum, orders that would otherwise overlap can be separated by orthogonal dispersions, as shown in Fig. 78, where a low-dispersion prism is perpendicular to a high-dispersion grating. Overlapping orders can be used for calibration in spectral regions where there are no reference lines.

At $\lambda = 0.5$ μm, a grating of width $W = 1$ cm has a maximum resolving power of 40,000 and a maximum resolution of 0.12 Å. If the grating period is $p = 2$ μm (500 grooves/mm), a normally incident beam has a first-order

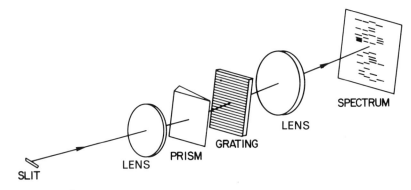

FIG. 78. A grating crossed with a prism of far lower dispersion to separate orders that would otherwise overlap.

diffraction angle of 15°, and the dispersion in this direction is about 1"/Å. The resolving power is 5000 and the resolution is 1 Å. For the central wavelength $\lambda = 0.5$ μm, the free spectral range is from 0.33 μm to 0.67 μm.

Large ruled gratings may have 600 to 1200 groves per millimeter over widths of 20 cm and resolving powers in the first order of 120,000 or 240,000. Echelle gratings (Fig. 6.23) have low periods (typically 30, 75, or 300 groves/mm) and operate in the Littrow mode at orders as high as 10,000, achieving resolutions as great as 10^6. Much higher resolving powers can be obtained with gratings than with prisms.

The instrument function for a transmission grating with equal clear and opaque strips is given in Eq. (6.42); and Fig. 17 in chapter 6 shows this function for a grating of five periods. The ratio of the space between orders to the central width of the peaks is the number of resolvable wavelengths within a free spectral range, which equals the number of periods across the grating, as found also in eq. (4.59). This instrument function was derived using scalar diffraction theory, which does not describe grating behavior exactly. Electromagnetic theory[184,185] is required to account for polarization effects and for the distribution of energy among the orders.

One phenomenon not explainable by scalar theory is *Wood's anomalies*,[186] abrupt changes in the power in an order as the incidence angle varies. The change in an order occurs when another order in the same wavelength is tangential to the grating surface, i.e., when $\theta' = 90°$.

A *blazed grating* has shaped grooves that concentrate energy in a particular order. The ratio of the energy in a given order to that specularly reflected from a blank of the same material is the *diffraction efficiency*. Efficiencies as high as 80% are possible, and blazed gratings can be more efficient than prisms. Reflection gratings are blazed by angling the groove faces so that the direction of specular reflection is the same as the diffraction direction in the desired order. For transmission gratings, the refraction angle is the same as that of diffraction. A blaze can be optimized for only a single wavelength in a given order, and a blaze for λ in the first order is also correct for $\lambda/2$ in the second, and so on. Efficiency is also increased for nearby wavelengths, and efficiencies in the first order can be improved from about 0.67λ to 2λ, which includes the entire free spectral range. A blaze optimized for higher orders increases efficiency to about 1.5λ. Efficiency depends in a complicated way on polarization, period, and groove shape. Loewen *et al.*[187] give a number of efficiency curves for the Littrow configuration.

The *concave grating*, invented by Rowland, has a spherical surface, and the projection of its grooves on a chord are equally spaced. Such a grating both disperses the light and images the slit, so no additional optics are required, which is a particular advantage at short wavelengths where optics may be very absorbing. The *Rowland circle* has a diameter that equals the radius of curvature of the grating and is tangent to the grating in a plane perpendicular to the grating lines. In all orders and for all wavelengths, points on the Rowland circle have their tangential astigmatic foci on the circle (Fig. 79). Thus a slit on the Rowland circle has an image that is

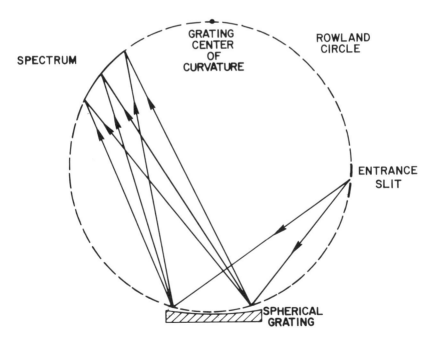

FIG. 79. The Rowland circle, whose diameter equals the radius of curvature of the spherical grating. Slits on the circle are imaged astigmatically on the circle as lines.

slitlike. However, the image irradiance is not uniform along its length. The length of the astigmatic foci increases with the height of the grating. Blaze angle cannot be optimized for even a single wavelength over the entire surface of a concave grating.

Most spectrometers with concave gratings make use of the properties of the Rowland circle. In the Paschen-Runge mounting, a large portion of the Rowland circle is used, so that much of the spectrum can be photographed in one exposure. The compact Eagle mount is a Littrow configuration. There are grazing incidence configurations useful at short wavelengths where reflectivities are low. The Wadsworth arrangement is not based on the Rowland circle; the incoming light is collimated, and the outgoing order is focused along the central grating normal, giving an image without astigmatism.

Most scanning grating spectrometers use plane gratings and only reflective components, and so focus adjustment for chromatic aberration is not required. One of the most common systems is the in-plane Ebert-Fastie arrangement (Fig. 80), in which the light from the entrance slit is collimated by a mirror, diffracted by the grating, and refocused. Large power throughput is achieved with entrance and exit slits located on a common circle, centered on the system axis, so that the astigmatic foci are tangential to the slits. In the off-plane version of this arrangement, the grating is rotated 90° about the system axis, and the slits are in the radial direction. The Czerney-Turner arrangement is similar but has separate mirrors for collimating and focusing. The Seya scanning spectrometer uses a concave grating.

The spectral properties of gratings have a number of applications outside of spectrometers proper. Gratings in Littrow mode serve as tuning mirrors at the end of the laser cavities.[188] Gratings can also be used as specularly

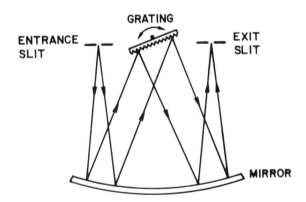

FIG. 80. The Ebert-Fastie arrangement. The spectrum is scanned by rotating the grating.

reflecting the rejection filters, if their period and blaze are such that unwanted wavelengths are preferentially diffracted from the zeroth order.

4.11.4. Fabry–Perot Spectrometers

The spectral unit of the Fabry-Perot spectrometer is the etalon, which is discussed in Chapter 5. The spectrometer and other applications of the etalon, such as wavelength measurement, are discussed by Candler,[136] Tolansky,[189,190] Jacquinot,[191] James and Sternberg,[139] Cook,[192] Thorne,[141] and Steel.[193] The Fabry-Perot in astronomy is discussed by Roesler.[194]

The etalon is a multiple-beam device, which produces by partial reflections a series of virtual sources with a longitudinal geometrical separation equaling twice the distance between the reflecting surface. The order of interference is the number of wavelengths between successive passes through the cavity. For a beam making an angle between the plates of θ from the plate normal, the order is

$$M = 2n\frac{s}{\lambda}\cos\theta = 2ns\sigma\cos\theta, \qquad (4.61)$$

where s is the plate separation, λ is the vacuum wavelength, and n is the refractive index of the medium between the plates. For a given separation, $\theta = 0$ gives the maximum order, $M_0 = 2sn/\lambda$. Equation (4.61) must be modified if there is a phase shift, other than 180°, introduced upon reflection. The phase shift and the magnitude of the reflectivity can depend on wavelength and angle, but we treat only the simple case in which they are constant.

The inside surfaces of an etalon are ideally flat and parallel. (The plates may be slightly wedged so that secondary reflections from the outer surfaces do not interfere with light from the inner surfaces.) If both inner surfaces have an irradiance reflectance R, transmittance T, and absorptance B, where

$$1 = R + T + B, \qquad (4.62)$$

then the transmittance of the etalon, written as a function of interference order, is

$$H(M) = \frac{T^2}{1 + R^2 - 2R\cos(2\pi M)} = \frac{T^2}{(1-R)^2} \frac{1}{1 + \left(2\frac{\mathscr{F}}{\pi}\right)^2 \sin^2(\pi M)}.$$
(4.63)

This is called the *Airy function* and is shown in Fig. 28, Chapter 1,

Volume 26. The *finesse* is

$$\mathcal{F} = \frac{\pi\sqrt{R}}{1-R}. \tag{4.64}$$

Another quantity sometimes used is the *coefficient of finesse*, $2\mathcal{F}/\pi$. For $R > 0.8$ or $\mathcal{F} > 10$, the reflectivity can be found from the finesse to an accuracy of better than 3% by the expression

$$R \sim 1 - \frac{\pi}{\mathcal{F}}. \tag{4.65}$$

The angular separation of the centers of the instrument function peaks depends on the order number, and thus, on the plate separation. The angular width of the peaks depends on the reflectivity of the plates, i.e., on the finesse. Differentiating Eq. (4.61) gives the angular dispersion of an etalon, $d\theta/d\lambda = M/2ns \sin\theta = \cot\theta/\lambda$.

The reflectivity determines the ratio of the magnitudes of the successive contributions to the etalon output and, hence, the shape of the instrument function. For an etalon used in transmission, all the contributions to the output pass through each surface once, so that the surface transmittances affect only the overall magnitude of the output. The ratio of the maximum transmittance of the etalon to the minimum is $(1+R)^2/(1-R)^2 \sim 4\mathcal{F}^2/\pi^2$. This ratio determines whether faint spectral lines can be detected in the presence of strong lines. The maximum transmissivity is

$$H_{\max} = \frac{T^2}{(1-R)^2} = \frac{1}{\left(1+\frac{B}{T}\right)^2} = \left(1-\frac{B}{1-R}\right)^2. \tag{4.66}$$

The fraction of the light transmitted over a full period of $H(M)$ is $T^2/(1-R^2)$, which equals $(1-R)/(1+R) \sim \pi/2\mathcal{F}$ if there is no absorption.

The resolution of a Fabry–Perot etalon is usually taken to be the full width of the instrument function at half maximum, which is

$$\delta\lambda = \frac{\lambda}{M_0\mathcal{F}} \quad \text{or} \quad \delta\sigma = \frac{1}{2ns\mathcal{F}}. \tag{4.67}$$

For lines of identical power so separated, the central irradiance of the intensity is about 88% of its maximum. The minimum separation for which there is any dip at all, the Sparrow resolution criterion, is about half that given by Eq. (4.67). The resolving power associated with the resolution of Eq. (4.67) is

$$\mathcal{R} = M_0\mathcal{F}. \tag{4.68}$$

The free spectral range, the separation between adjacent peaks of the instrument function, is

$$\Delta\lambda = \frac{\lambda}{M} = \frac{\lambda^2}{2ns\cos\theta} \quad \text{or} \quad \Delta\sigma = \frac{\sigma}{M} = \frac{1}{2ns\cos\theta} = \mathscr{F}\delta\sigma. \quad (4.69)$$

The second of these equations shows that the finesse equals the number of resolvable elements within the free spectral range, $\Delta\sigma/\delta\sigma$. The effective number of beams of an etalon is taken to equal its finesse by analogy to a grating [comparing Eq. (4.55) and Eq. (4.68)], for which the number of beams equals the number of periods. The greater the reflectivity and the finesse, the slower the drop in the intensity of successive virtual sources. Thus more source images interfere to give the net intensity, and the optical path length between the first virtual source and that of any relative intensity is greater.

An etalon by itself is an angular filter, and a spectrometer requires a focusing lens to produce a spatial pattern. (The parallel etalon is said to produce fringes at infinity.) In the traditional spectrograph configuration (Fig. 81), the output for a single wavelength is a series of rings, whose diameters are approximately

$$d' = f'\sqrt{8\left(1 - \frac{M}{M_0}\right)}, \quad (4.70)$$

where f' is the focal length of the focusing lens. the illumination arrangement does not affect the shape of these fringes, which modulate whatever irradiance would be present in the focal plane of the output lens if the etalon were not present. Input optics are usually required to produce a nominally uniform output, which is desirable in a spectrograph. If the output field is

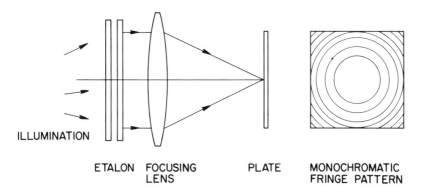

ILLUMINATION

ETALON   FOCUSING   PLATE   MONOCHROMATIC
         LENS   FRINGE PATTERN

FIG. 81. The Fabry-Perot spectrograph. A series of rings is formed by the light at each wavelength.

uniform without the etalon in place, then all the rings produced by the etalon contain the same amount of energy, since their widths are inversely proportional to their diameters. The diameter of the half maximum of the central fringe is found by setting $M = M_0 - 1/\mathscr{F}$ in Eq. (4.70) and using Eq. (4.67) to give

$$d' = f'\sqrt{\frac{8}{\mathscr{R}}}. \tag{4.71}$$

Taking the energy in this diameter as the spectrometer throughput, the luminosity-resolving power product for an etalon of area A is

$$\mathscr{L} \times \mathscr{R} = 2\pi A \tau. \tag{4.72}$$

For a given resolution, the throughput of an etalon spectrometer is several hundred times that of a spectrometer using a grating of equal area. The spherical Fabry-Perot (Chapter 1, Volume 26) is not governed by Eq. (4.72). Its order number is, to a first approximation, independent of θ, so its throughput is proportional to its resolution.

The ratio of the area under the peaks of the instrument function to that of the entire instrument function increases with reflectivity. In the limit as $R \to \infty$, the energy between the half maxima is 0.7 of the total energy.

The performance of an etalon depends critically on the flatness of its surfaces and coatings. The actual instrument function is the convolution of the ideal instrument function with a function that accounts for degradation. The width of the degradation function depends on the ratio of the surface variations to the wavelength. In practice, surface roughness sets an upper limit on the finesse. If the RMS roughness is $\lambda/50$, the finesse cannot exceed about 25.

A perfect etalon with surface reflectivities of 0.9 has a finesse of 30. If the cavity is air-filled, with a separation of 1 mm, at $\lambda = 0.5$ μm, the central order of interference is 4000, and order 3999 is at an angle of 1.3°. The resolving power is 120,000, the resolution is 0.04 Å, and the free spectral range is 1.25 Å. The maximum of the instrument function is 400 times the minimum. For a focusing lens of 100 mm focal length, the first four ring diameters are 0, 4.5 mm, 6.3 mm, and 7.8 mm, and the diameter at the half maximum of the central spot is 75 μm. If the reflectivity is increased to 0.95, the finesse is increased to 61, and the resolution is doubled. Doubling the plate separation also gives twice the resolution, but half the free spectral range. With no absorption, 5% of the incident energy is in the output. With 5% absorption and 90% reflectivity, the maxima drop to 25%, and the total energy in the fringe pattern is 1% of that incident.

The optimum reflective coating depends upon the application. For metallic films, both absorption and reflectivity increase with film thickness, so

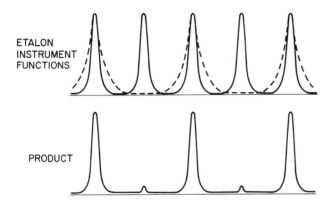

FIG. 82. The product of two etalon instrument functions, showing how a pair of etalons in series can have increased free spectral range and somewhat better resolution. The two etalons here have a spacing ratio of 2:1.

resolution and throughput are opposed. An aluminum film can have a reflectivity of around 85% with little phase variation, and silver can give about 90% reflectivity with less absorption. Dielectric films absorb less than metal films, but they usually cause wavelength-dependent phase shifts. Therefore, dielectrics are useful with small wavelength ranges, for example, in the determination of line shape, and metals are useful for large ranges, for example, in wavelength determination.

Because of its small free spectral range, the etalon is frequently used with a dispersion spectrometer, which decreases the luminosity. The free spectral range can be increased by using etalons in series, the ratio of whose spacings is that of two integers, so that some of the transmission maxima coincide. The instrument function is the product of the two instrument functions, as shown in Fig. 82 for a 2:1 spacing ratio. The free spectral range is that of the thinner etalon, and the resolution is somewhat greater than that of the thicker one, since the product of two instrument function peaks is narrower than either. If the spacing ratio is ρ, the resolving power is about $(1+\rho)^{1/2}$ that of the thicker etalon. The angular acceptance of the pair is that of the lesser of the two, the higher-resolution etalon.

The etalon can be used "on axis" as a monochromator or scanning spectrometer (Fig. 83). The instrument function of an on-axis etalon monochromator is found by convolving the function due to the etalon alone with a function that depends upon the angular subtense of the detector aperture. The wavelength at which maximum power reaches the detector is determined by the optical path length between plates, which can be varied by changing the plate separation or the refractive index between the plates.

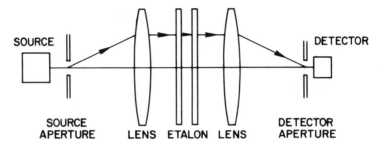

FIG. 83. On-axis Fabry–Perot arrangement, in which only the central order is detected. The wavelength of this order is varied by changing the optical path length between the etalon plates.

In *pressure scanning*, the index of a gas within the cavity is changed by varying its pressure. To change the order number from M to $M + \Delta M$ requires a pressure change (in atmospheres) of approximately

$$\Delta P = \frac{\Delta n}{\beta n} = \frac{\Delta M}{M}, \qquad (4.73)$$

where β depends on the gas. For air, $\beta \sim 2.7 \times 10^{-4}$. For an etalon with a 1 mm separation, a change of one order at $\lambda = 0.5$ μm requires a pressure change of about one atmosphere. Etalons in series enclosed in a common vessel can be pressure-scanned synchronously. Pressure cannot be measured to the accuracies required for spectroscopy, but light at a reference wavelength can be introduced to calibrate the scan.

Wavelength scanning is also done by moving one of the plates with piezoelectric or magnetostrictive devices, for instance. The plates must be kept parallel. A movement of only $\lambda/2$ is required to cover the free spectral range; so rapid scanning is possible, and scan times of a microsecond have been achieved.[195] Scanning etalons can be very rigid, since scanning can be done with only slight movements and with no moving parts. Scanning techniques can also be used to servo nominally fixed etalons and to adjust the ratio of spacings for etalons in series.

4.11.5. The Fourier Transform Spectrometer

Fourier transform spectroscopy, or simply Fourier spectroscopy, began with the work of Michelson,[196] who used his interferometer to determine the separation of the lines of the sodium doublet from the variation in fringe visibility with mirror displacement. Fourier spectroscopy was little used until the 1950s, when it was recognized that it made possible improved signal-to-noise ratio in the infrared. The development of digital computers made practical the required numerical processing. The later discovery of the fast Fourier transform algorithm (Brigham[197]) greatly reduced computation time. Most visible and near-infrared Fourier spectroscopy is done with derivatives of the Michelson interferometer, and we discuss only these instruments. Fourier spectroscopy is treated in several books (Mertz,[198] Bell,[199] Chamberlain,[200] Schnopper and Thompson[201]), review articles (Vanasse and Sakai,[202] Sakai,[203] Bell,[204] Breckinridge and Schindler[205]), and conference proceedings (Vanesse et al.[206]). The original work of J. Connes has been translated into English (Connes[207]). Bell[199] and Lovell[208] treat commercially available instruments.

The Fourier transform spectrometer can be most easily understood beginning with the Twyman–Green interferometer (Fig. 84), a form of the Michelson interferometer whose entrance aperture is sufficiently small that the light beyond the first lens is nearly collimated. Suppose that the mirrors are identical and parallel and that the beam splitter provides a 50% division with no wavelength-dependent phase shift. If the incoming light is monochromatic, with wave number σ, then as one of the mirrors is translated, the power reaching the detector is

$$P(x) \propto 1 + \cos(2\pi\sigma x). \quad (4.74)$$

Here x is the optical path difference between the two arms, which is twice the displacement of the mirror from the zero OPD position. With polychromatic light, the output power is the sum of independent contributions from

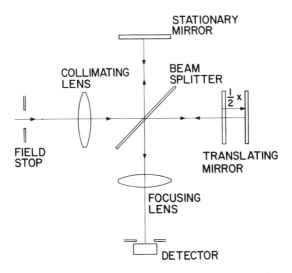

FIG. 84. The Michelson interferometer, the basis for most Fourier transform spectrometers.

each wavelength, each being modulated at a different rate with changing OPD. If $S(\sigma)$ is the spectrum, the output is

$$P(x) \propto \int_0^\infty d\sigma\, S(\sigma)[1+\cos(2\pi\sigma x)]$$

$$= \int_0^\infty d\sigma\, S(\sigma) + \int_0^\infty d\sigma\, S(\sigma)\cos(2\pi\sigma x). \qquad (4.75)$$

This function is symmetrical about the zero OPD position. A record of the output as a function of x is called an *interferogram*. Examples of noise-free interferograms are shown in Fig. 85 for (*a*) monochromatic light, (*b*) a single spectral line of finite width, (*c*) a pair of narrow lines of identical strength, and (*d*) a continuous source with a nonconstant spectrum. The output consists of two parts. The constant term $\int S(\sigma)d\sigma$ depends on the total power. The spectral information is contained in the varying term

$$F(x) = \int_0^\infty d\sigma\, S(\sigma)\cos(2\pi\sigma x), \qquad (4.76)$$

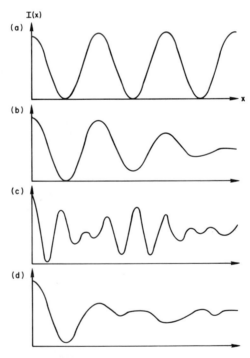

FIG. 85. Typical Fourier spectrometer outputs for light that is (a) monochromatic, (b) has a single wide spectral line, (c) has a pair of wide lines, (d) has an arbitrary spectrum.

called the *interferogram function*. The shape of the spectrum is the cosine transform of the interferogram function,

$$S(\sigma) \propto \int_0^\infty dx\, F(x) \cos(2\pi\sigma x). \quad (4.77)$$

This equation requires that $F(x)$ be known for all x, which would require an infinite translation of the interferometer mirror. Let X be the maximum path difference obtained by the interferometer. If the measured interferogram is taken to vanish for path lengths greater than X, then this function is the product of the "true" interferogram and a rectangle function, which equals unity for $|x| \leq X$ and vanishes for $|x| > X$. So the calculated spectrum $S'(\sigma)$ is related to the true spectrum $S(\sigma)$ by the convolution

$$S'(\sigma) = S(\sigma) \otimes \left[2X \frac{\sin(2\pi\sigma X)}{2\pi\sigma X} \right]. \quad (4.78)$$

Thus the normalized instrument function is

$$H(\sigma) = \mathrm{sinc}(2\sigma X) = \frac{\sin(2\pi\sigma X)}{2\pi\sigma X}. \quad (4.79)$$

This function is not positive definite, unlike that for other spectrometers in which the instrument function represents a power measurement. By the Rayleigh resolution criterion, lines can be distinguished that are as close as about

$$\delta\lambda = \frac{\lambda^2}{2X} \quad \text{or} \quad \delta\sigma = \frac{1}{2X}. \quad (4.80)$$

So the resolving power is

$$\mathcal{R} = \frac{2X}{\lambda} = 2\sigma X, \quad (4.81)$$

which equals twice the number of modulations undergone by the contribution due to wavelength λ as the OPD varies by X. Just-resolved adjacent wavelengths are modulated $\mathcal{R} \pm 1/2$ times.

If the zero OPD position is not known accurately, an erroneous spectrum is calculated by using the cosine transform. This difficulty is overcome by recording a nominally symmetric two-sided interferogram, whose Fourier transform is complex. The magnitude of its imaginary part depends upon the asymmetry of the motion, and it can be removed mathematically.

For a maximum OPD of $X = 1$ cm, the resolution for all wave numbers is $\delta\sigma = 0.5/$cm. At $\lambda = 1.0$ μm, the wavelength resolution is $\delta\lambda = 0.008$ Å, and the resolving power is 20,000.

Instrument functions and computed spectra can be modified numerically by a process called *apodization*. The measured interferogram is multiplied by the so-called apodization function, so the instrument function becomes the convolution of the function of Eq. (4.79) with the Fourier transform of the apodization function. To change the computed spectrum, either the interferogram can be multiplied and transformed, or the spectrum computed without apodization can be convolved with the transform of the apodization function. The apodization function is chosen to remove misleading artifacts from the computed spectrum, typically by requiring that the instrument function be positive definite or "less bumpy." For example, a triangle apodization gives an instrument function $\text{sinc}^2(2\sigma x)$, like that of a grating. In general, the instrument function is widened as it is smoothed. Other applications of apodization, including its use in imaging systems, are discussed by Jacquinot and Roizen-Dossier.[209]

$P(x)$ cannot be measured at all points, but must be sampled. According to the well-known result from sampling theory (Bracewell[210]), measurements must be made at a rate of at least twice the highest modulation frequency component in the output. That is, if λ_{min} is the shortest wavelength in the spectrum, the intensity sample must be sampled for every OPD change of $\lambda_{min}/2$, or for every $\lambda_{min}/4$ of mirror displacement. If the maximum wave number is σ_{max}, the required number of measurements is

$$N = \frac{2X}{\lambda_{min}} = 2X\sigma_{max} = \frac{\sigma_{max}}{\delta\sigma}, \qquad (4.82)$$

which equals the number of resolvable spectral elements. The values of x at which $P(x)$ should be sampled must be accurate to a small fraction of the shortest wavelength. This requirement is one of the factors determining the minimum wavelength at which Fourier spectrometers can operate. Unequal sampling intervals give errors analogous to those due to gratings with irregular line spacings (Guelachvili,[211] Zachor et al.[212]). Laser interferometers are commonly used to monitor mirror position.

Unless the source is sufficiently bright, the entrance aperture must be large enough so that the light within the interferometer is composed of beams with a finite range of angles. The optical path difference for a beam varies as the cosine of its angle from the mirror normal, so that the entire output field is not modulated in phase. Away from the axis of the focusing lens, the modulation rate decreases; so bulls-eye patterns appear and spread as the path difference is increased. The central fringe diameter decreases with OPD, and therefore its minimum size decreases with resolving power. The detector must be sufficiently small to cover only a region of the interference pattern where the irradiance is no more than 180° out of phase with that on axis. If Ω' is the solid angle subtended by the detector, this restriction gives

$$\Omega'\mathcal{R} = 2\pi. \qquad (4.83)$$

Multiplying this equation by the mirror area gives the same $\mathcal{L} \times \mathcal{R}$ product as the Fabry-Perot Eq. (4.72). Because of finite detector size, the measured modulation rate is less than that at the center of the pattern, a correction which must be made in the inversion equation.

The form of the interference pattern changes with the mirror translation, because in addition to changing the OPD, one of the entrance aperture images is translated. *Field-widened* interferometers[193,213] have fixed entrance aperture images, and greater $\mathcal{L} \times \mathcal{R}$ factors. The system shown in Fig. 86, due to Bouchareine and Connes,[214] illustrates one approach, namely the use of dielectric compensators. A prism with a mirrored rear surface translates tangentially to the geometrical image of the mirror, so that the position of the entrance aperture image is fixed. However, the changing thickness of glass through which the light passes gives an OPD change, represented in the figure by the "optical path length mirror." The other arm contains an identical prism to compensate for dispersion. This instrument is limited by aberrations. In another type of field-widening technique, in each arm there is an afocal system, one of which is translated to change the OPD without shifting the entrance aperture image.

The mechanical motion in the spectrometer must be smooth and rigid to preserve alignment during the scan. Even slight tilts change the fringe

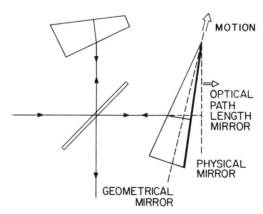

FIG. 86. A field-widened Fourier spectrometer, in which the optical path length is changed without moving the geometrical image of the source. Thus the shape of the bull's-eye fringes does not change with OPD, except for expanding or contracting. The path length is changed by translating a prism coated with a reflecting surface parallel to the image of the surface, so that the geometrical image of the mirror does not move, although the thickness of glass traversed by the light changes.

pattern, giving spurious interferograms. A number of tilt-insensitive arrangements have been devised, using corner reflectors, cat's eyes, and so on. In another approach, both mirrors are located on the same carriage, and the light paths are arranged so that tilts affect both path lengths identically.

The beam splitter is a critical element. It may introduce polarization effects that decrease the output modulation, and its behavior may change with wavelength. If there are wavelength-dependent phase shifts, then there is no unique OPD zero, and so a compensator is required.

A Michelson interferometer can also be used to measure the refractive index of a transmitting sample placed in one of its arms. The index of a reflective sample is found by using it as one of the mirrors.

Through the entire scan, the detector in a Michelson spectrometer receives radiation from all wavelengths, so the Fellgett advantage is realized. This is not the case for all Fourier spectrometers.

The Fourier spectrometers suffer from unique forms of noise, and the references should be consulted for a discussion of this subject. One way to obtain a high signal-to-noise ratio is by averaging the results from a number of scans. Some instruments can scan in less than a second, and signal-to-noise values of 10^6 have been reported.

References

Introduction

1. R. Kingslake, ed., *Applied Optics and Optical Engineering*, Vol. 4, Optical Instruments, Part 1, Academic Press, New York, 1967.
2. R. Kingslake, ed., *Applied Optics and Optical Engineering*, Vol. 5, Optical Instruments, Part 2, Academic Press, New York, 1969.
3. R. Kingslake, *Optical System Design*, Academic Press, New York, 1983.
4. D. H. Horne, *Optical Instruments and Their Applications*, Adam Hilger, Bristol, 1980.
5. R. Glazebrook, *A Dictionary of Applied Physics*, Macmillan, New York, 1922.
6. R. Kingslake, *Lenses in Photography*, A. S. Barnes, New York, 1963.
7. A. Cox, *Photographic Optics*, Focal Press, London, 1966.

The Visual System

8. G. A. Fry, "The Eye and Vision," in *Applied Optics and Optical Engineering*, Vol. 2, p. 1 (R. Kingslake, ed.), Academic Press, New York, 1965.
9. G. A. Fry, "The Optical Performance of the Human Eye," in *Progress in Optics*, Vol. 7, Chap. 1 (E. Wolf, ed.), North-Holland Publ., New York, 1970.
10. M. Alpern, "The Eyes and Vision," in *Handbook of Optics*, (W. G. Driscoll, ed.), Chap. 10, McGraw-Hill, New York, 1978.
11. C. H. Graham, ed. *Vision and Visual Perception*, Wiley, New York, 1965.
12. M. H. Pirene, *Vision and the Eye*, Chapman and Hall, London, 1967.

13. T. N. Cornsweet, *Visual Perception*, Academic Press, New York, 1970.
14. G. S. Brindley, *Physiology of the Retina and Visual Pathway*, Edward Arnold, London, 1970.
15. J. W. T. Walsh, *Photometry*, Constable, London, 1958.
16. J. M. Bennett and H. W. Bennett "Polarization," in *Handbook of Optics* (W. G. Driscoll, ed.), Chap. 10, McGraw-Hill, New York, 1978.
17. L. A. Riggs, "Visual Acuity," in *Vision and Visual Perception* (C. H. Graham, ed.), Wiley, New York, 1965.

Magnifiers and Eyepieces

18. G. A. Boutry, *Instrumental Optics* (R. Auerbach, tranl.), Interscience Publishers, New York, 1962.
19. L. Rosin, "Eyepieces and Magnifiers," in *Applied Optics and Optical Engineering*, Vol. 3, Chap. 3 (R. Kingslake, ed.), Academic Press, New York, 1965.
20. R. Kingslake, *Lens Design Fundamentals*, Academic Press, New York, 1965.
21. R. C. Jones, "Immersed Radiation Detectors," *Appl. Opt.* 1, 607 (1962).

Afocal Systems

* Reference added in proof: W. B. Wetherell "Afocal lenses," in *Applied Optics and Optical Engineering*, Vol. 10, (R. R. Shannon, J. C. Wyart eds.), Academic Press, New York, 1987.
22. R. E. Hopkins, "Mirror and Prism System," in *Applied Optics and Optical Engineering*, Vol. 3, p. 269 (R. Kingslake, ed.), Academic Press, New York, 1965.
23. D. W. Swift, "Image Rotation Devices—A Comparative Survey," *Opt. and Laser Tech.* 4, 175 (1972).
24. J. Dyson, "Unit Magnification Optical System Without Seidel Aberrations," *J. Opt. Soc. Am.* 49, 713 (1959).
25. C. G. Wynne, "A Unit Power Telescope for Projection Copying," in *Optical Instruments and Techniques* (J. H. Dickson,), Oriel Press, Newcastle-upon-Tyne, England, 1969.
26. A. Offner, "New Concepts in Projection Mask Aligners," *Opt. Eng.* 14, 130 (1975).
27. B. J. Lin, "Optical Methods for Fine Line Lithography," in *Fine Line Lithography* (R. Newman, ed.), Chap. 2, North-Holland Publ., Amsterdam, 1980.
28. J. M. Burch and and D. C. Williams, "Afocal Lens Systems for Straightness Measurements," *Opt. and Laser Tech.* 6, 166 (1974).
29. J. W. Goodman, *Introduction to Fourier Optics*, McGraw-Hill, New York, 1968.

Autocollimators

30. K. J. Hume, *Metrology With Autocollimators*, Hilger and Watts, London, 1965.
31. J. A. Mauro, ed., *Optical Engineering Handbook*, General Electric, Syracuse, New York, 1966.
32. A. W. Young, "Optical Workshop Instruments," in *Applied Optics and Optical Engineering*, Vol. 4, Chap. 4 (R. Kingslake, ed.), Academic Press, New York, 1967.
33. K. J. Habell and A. Cox, *Engineering Optics*, Pitman and Sons, London, 1953.
34. H. E. Torberg, W. J. Rowan, and J. E. Vyce, "Optical Instruments for Metrology," Chap. 16, in *Handbook of Optics* (W. G. Driscoll, ed.), McGraw-Hill, New York, 1978.
35. S. C. Bottomley, "Recent Developments in Autocollimating Telescopes," *Precision Engineering* 3, 149 (1981).

Projectors

36. H. H. Hopkins, "On the Diffraction Theory of Optical Images," *Proc. Roy. Soc.* (London) **A217**, 408 (1953).
37. H. H. Hopkins, "Applications of Coherence Theory in Microscopy and Interferometry," *J. Opt. Soc. Am.* **47**, 508 (1957).
38. B. J. Thompson, "Image Formation with Partially Coherent Light," in *Progress in Optics*, Vol. 7, p. 171 (E. Wolf, ed.), North-Holland, Publ. Amsterdam, 1969.
39. M. Born and E. Wolf, *Principles of Optics*, Pergamon Press, Oxford, 1975.
40. A. S. Marathay, *Elements of Optical Coherence Theory*, Wiley, New York, 1981.
41. K. Singh and M. De, "Bibliographic Review on Studies in Partial Coherence," *Journal of Optics* (India), in three parts, **5**, 42 (1976); **6**, 15 (1978); **6**, 77 (1978).
42. J. W. Goodman, *Statistical Optics*, Wiley, New York, 1985.
43. D. S. Goodman, *Stationary Optical Projectors*, Ph. D. Thesis, The University of Arizona, Tucson, 1979.
44. L. C. Martin, *The Theory of the Microscope*, American Elsevier, New York, 1966.
45. L. Rayleigh, "On the Theory of Optical Images With Special Reference to the Microscope," *Phil. Mag.* V **42**, 167 (1896).
46. H. H. Hopkins and P. M. Barham, "The Influence of the Condenser on Microscope Resolution," *Proc. Roy. Soc.* **370B**, 737 (1951).
47. D. N. Grimes and B. J. Thompson, "Two-Point Resolution with Partially Coherent Light," *J. Opt. Soc. Am.* **57**, 1330 (1967).
48. D. N. Grimes, "Imaging of Tri-Bar Targets and the Theoretical Resolution Limit in Partially Coherent Illumination," *J. Opt. Soc. Am.* **61**, 870 (1971).
49. B. M. Watrasiewics, "Theoretical Calculations of Images of Straight Edges in Partially Coherent Illumination," *Optica Acta* **12**, 391 (1965).
50. E. C. Kintner and R. M. Sillitto, "Edge Ringing in Partially Coherent Imaging," *Optica Acta* **24**, 591 (1977).
51. D. A. Tichenor and J. W. Goodman, "Coherent Transfer Function," *J. Opt. Soc. Am.* **62**, 293 (1972).
52. D. S. Goodman, "Condenser Aberrations in Koehler Illumination," *J. Opt. Soc. Am. A* **1**, 1316 (1984). (Abstract only.)
53. F. Zernike, "The Concept of Degree of Coherence and its Application to Optical Problems," *Physica* **5**, 785 (1938).
54. M. Lacombat, G. M. Dubroeucq, J. Massin, and M. Brebignon, "Laser Projection Printing," *Solid State Technology* **23**, 115 (1979).

Microscopes

55. J. R. Benford and H. E. Rosenberger, "Microscopes," in *Applied Optics and Optical Engineering*, Vol. 4, p. 31 (R. Kingslake, ed.), Academic Press, New York, 1967.
56. W. Burrells, *Microscope Techniques*, Wiley, New York, 1977.
57. G. H. Needham, *The Practical Use of the Microscope*, Charles C. Thomas, Springfield, Ill., 1977.
58. E. M. Chamot, and C. W. Mason, *Handbook of Chemical Microscopy*, Vol. 1, Wiley, New York, 1958.
59. M. Francon, *Progress in Microscopy*, Row, Peterson and Company, Evanston, Ill., 1961.
60. H. C. Claussen, "Mikroskope," in *Handbuch der Physik*, Vol. 29, p. 343 (S. Flügge, ed.), Springer-Verlag, Berlin, 1967.

61. P. Gray, ed., *The Encyclopedia of Microscopy and Microtechnique*, Van Nostrand Reinhold, New York, 1973.
62. J. R. Meyer-Arendt, "Microscopy as a Spatial Filtering Process," in *Advances in Optical and Electron Microscopy*, Hilger and Watts, London, 1964.
63. J. R. Benford, "Microscope Objectives," in *Applied Optics and Optical Engineering*, Vol. 3, Chap. 3 (R. Kingslake, ed.), Academic Press, New York, 1967.
64. J. R. Benford and H. E. Rosenberger, "Microscope Objectives and Eyepieces," in *The Handbook of Optics*, Chap. 6 (W. G. Driscoll, ed.), McGraw-Hill, New York, 1978.
65. J. G. Delly and W. F. Wills, "How to Buy a Compound Microscope," p. 66, *American Laboratory*, April (1965).
66. J. R. Meyer-Arendt, "Optical Instrumentation for the Biologist: Microscopy," *Appl. Opt.* **4**, 1 (1965).
67. P. E. Ciddor, "Photoelectric Setting Microscopes," in *Advances in Optical and Electron Microscopy* Vol. 8, p. 25 (R. Barer, V. E. Cosslett, eds.), Academic Press, New York, 1982.
68. D. H. Freeman, "Measurement of Microscope Magnification," *Appl. Opt.* **3**, 1005 (1964).
69. G. W. W. Stevens, *Microphotography*, Interscience Publishers, New York, 1960.
70. K. P. Norris, "Development of Reflecting Microscopes," in *Modern Methods of Microscopy* (A. E. J. Vickers, ed.), p. 15, Butterworths Scientific Publications, London, 1956.
71. J. Rheinberg, "On an Addition of the Methods of Microscopical Research by a New Way of Optically Producing Colour-Contrast between an Object and its Background, or between Definite Parts of the Object itself," *Journal of the Royal Microscopical Society*, 373 (1896).
72. A. H. Bennett, H. Osterberg, H. Jupnik, and O. W. Richards, *Phase Microscopy*, Wiley, New York, 1951.
73. S. Tolansky, *Surface Microtopography*, Interscience Publishers, New York, 1960.
74. W. Krug, J. Rienitz, and G. Schultz, *Contributions to Interference Microscopy* (J. H. Dickson, trans.), Hilger and Watts, London, 1964.
75. N. H. Hartshorne and A. Stuart, *Crystals and the Polarizing Microscope*, Edward Arnold, London, 1950.
76. M. R. Young, "Principles and Techniques of Fluorescence Microscopy," *Quarterly Journal of Microscopical Science* **102**, 419 (1961).
77. T. Wilson, "Imaging Properties and Applications of Scanning Optical Microscopes," *Appl. Phys.* **22**, 119 (1980).
78. E. Ash, ed., *Scanned Image Microscopy*, Academic Press, New York, 1980.
79. C. J. R. Shepard, "Scanning Optical Microscope," *Electronics and Power* **26**, 166 (1980).
80. T. Wilson and C. J. R. Shepard, "Theory and Practice of Scanning Optical Microscopes," *Appl. Phys.* **22**, 119 (1980).
81. F. Roberts and J. Z. Young, "The Flying-Spot Microscope," *Proc. Inst. Elect. Eng.* **99**, 747 (1952).
82. W. J. Alford, R. D. Vanderneut, and V. J. Zaleckas, "Laser Scanning Microscopy," *Proc. IEEE* **70**, 641 (1982).
83. G. Bouwhuis and J. J. M. Braat, "Recording and Reading of Information on Optical Disks," in *Applied Optics and Optical Engineering*, Vol. 9, p. 73 (R. R. Shannon and J. C. Wyant, eds.), Academic Press, New York, 1969.
84. G. Bouwhuis, J. Braat, A. Huijser, J. Pasman, G. van Rosemalen, and K. Schouhamer Immink, *Principles of Optical Disc Systems*, Adam Hilger, Bristol, 1985.
85. J. Isailovic, *Videodisc and Optical Memory Systems*, Prentice-Hall, Englewood Cliffs, New Jersey, 1985.
86. R. E. Swing, "Microdensitometer Optical Performance: Scalar Theory and Experiment," *Opt. Eng.* **15**, 559 (1976).

87. J. B. De Velis, ed., Opt. Eng., November–December, 12 (1973). Entire issue on microdensitometry.

Schlieren Instruments

88. L. M. Foucault, "Description des Procédés Employés pour Reconnaître la Configuration des Surfaces Optiques, *Compt. Rend. Acad. Sci. Paris* **47**, 958 (1858).
89. A. Toepler, *Beobachtungen Nach Einer Neuen Optischen Methode, Poeggendorf's Ann. Phy. chem.* **127**, 556 (1986).
90. L. A. Vasil'ev, *Schlieren Methods*, Israel Program for Scientific Translations, New York, Jerusalem, London, 1971. Translated from the Russian *Tenevye Metody*, Izdatel'stvo "Nauka," Moscow, 1968.
91. H. Wolter, "Schlieren-, Phasenkontrast und Lichtschnittverfahren," in *Handbuch der Physik*, Vol. 24: Grundlagen der Optik, Springer-Verlag, Berlin, 1956.
92. W. Merzkirch, *Flow Visualization*, Academic Press, New York, 1974.
93. J. Ojeda-Castañeda, "Foucault, Wire, and Phase Modulation Tests," in *Optical Shop Testing* (D. Malacara, ed.), Wiley, New York, 1978.
94. R. J. North, "Schlieren Photography and Allied Techniques," in *SPSE Handbook of Photographic Science and Engineering* (W. Thomas, ed.), p. 1162, Wiley, New York, 1973.
95. J. W. Beams, "Shadow and Schlieren Methods," in *Physical Methods in Gas Dynamics and Combustion* (R. Landenburg, ed.), p. 26, Princeton University Press, Princeton, New Jersey, 1954.
96. D. W. Holder and R. J. North, "Schlieren Methods," Note on *Applied Science No. 31*, National Physical Laboratory, Her Majesty's Stationery Office, London, 1963.
97. N. F. Barnes and S. L. Bellinger, "Schlieren and Shadowgraph Equipment for Air Flow Analysis," *J. Opt. Soc. Am.* **35**, 497 (1945).
98. E. H. Linfoot, *Recent Advances in Optics*, Clarendon, Oxford, 1955.
99. P. Lloyd, *Optical Methods in Ultracentrifugation, Electrophoresis, and Diffusion*, Clarendon, London, 1974.
100. C. J. van Oss, "Methods Used in the Visualization of Concentration Gradients," in *Techniques of Surface and Colloid Chemistry and Physics*, Vol. 1, p. 213 (R. J. Good, R. R. Stromberg, and R. L. Patrick, eds.), Marcel Dekker, New York, 1972.
101. F. Zernike, "Phase Contrast, A New Method for the Microscopic Observation of Transparent Objects," *Physica* **9**, 686 (1942).
102. B. Lyot, Procédés Permettant d'Étudier les Irregularités d'une Surface Optique Bien Polie," *Compt. Rend. Acad. Sci.* (Paris) **222**, 765 (1946).
103. W. T. Welford, "Bubble Chamber Optics," *Appl. Opt.* **2**, 981 (1963).

Ellipsometers

104. R. M. A. Azzam and N. M. Bashara, *Ellipsometry and Polarized Light*, North-Holland Publ., Amsterdam, 1977.
105. E. Passaglia, R. R. Stromberg, and J. Kruger, eds., *Ellipsometry in the Measurement of Surfaces and Thin Films 1963 Symposium Proceedings*, Nat. Bur. Stds. Miscellaneous Publication W256, 1964.
106. N. M. Bashara, A. B. Buckman, and A. C. Hall, eds., *Recent Developments in Ellipsometry*, North-Holland Publ., Amsterdam, 1969.
107. N. M. Bashara and R. M. A. Azzam, eds., *Ellipsometry. Proceedings of the Third International Conference on Ellipsometry*, North-Holland Publ., Amsterdam, 1976.

108. R. H. Muller, R. M. A. Azzam, and D. E. Aspnes, eds., *Ellipsometry. Proceedings of the Fourth International Conference on Ellipsometry*, North-Holland Publ., Amsterdam, 1980.
109. W. L. Hyde and R. M. A. Azzam, eds., *Polarized Light. Proc. Soc. Phot. Opt. Instr. Eng.* **88**, (1976).
110. R. M. A. Azzam and D. L. Coffeen, eds., *Optical Polarimetry. Proc. Soc. Phot. Opt. Instr. Eng.* **112**, (1977).
111. A. C. Hall, "A Century of Ellipsometry," *Surface Science* **16**, 1 (1969).
112. A. Rothen, "Measurements of the Thickness of Thin Films by Optical Means, from Rayleigh and Drude to Langmuir, and the Development of the Present Ellipsometer," from *Ellipsometry in the Measurement of Surfaces and Thin Films*, 1963 Symposium N. B. S., Miscellaneous publication 256.
113. P. S. Hauge, "Survey of Methods for the Complete Determination of a State of Polarization," in *Polarized Light: Instruments, Devices and Applications* (W. L. Hyde and R. M. A. Azzam, eds.), *Proc. Soc. Phot. Opt. Instr. Eng.* **88**, 3 (1976).
114. P. S. Hauge, "Techniques of Measurement of the Polarization Altering Properties of Linear Optical Systems," in *Optical Polarimetry: Instruments and Applications* (R. M. A. Azzam and D. L. Coffew, eds.), *Proc. Soc. Opt. Instr. Eng.* **112**, 2 (1977).
115. P. S. Hauge, "Recent Developments in Instrumentation in Ellipsometry," *Surface Science* **96**, 108 (1980).
116. Z. Sekera, "Recent Developments in the Study of the Polarization of Sky Light," in *Advances in Geophysics*, Vol. 3, p. 43, Academic Press, New York, 1956.
117. R. M. A. Azzam, "Arrangement of Four Photodetectors for Measuring the State of Polarization of Light," *Opt. Lett.* **10**, 309 (1985).
118. D. E. Aspnes, "Spectroscopic Ellipsometry of Solids," in *Optical Properties of Solids, New Developments* (B. O. Seraphim, ed.), p. 799, American Elsevier, New York, 1976.

Refractomers

119. G. E. Fishter, "Refractometry," in *Applied Optics and Optical Engineering*, Vol. 4, p. 363 (R. Kingslake, ed.), Academic Press, New York, 1967.
120. R. S. Longhurst, *Geometrical and Physical Optics*, Chap. 5, Longmans, London, 1973.
121. A. J. Werner, "Methods in High Precision Refractomy of Optical Glasses," *Appl. Opt.* **7**, 837 (1968).
122. T. R. P. Gibb, *Optical Methods of Chemical Analysis* McGraw-Hill, New York, 1942.
123. F. E. Jones, "The Refractive Index of Air," *J. Res. Nat. Bur. Std.* **86**, 27 (1981).
124. L. W. Tilton, "Prism Refractometry and Certain Goniometrical Requirements for Precision," *J. Res. Nat. Bur. Std.* **2**, 909 (1929).
125. L. W. Tilton, "Prism Size and Orientation in Minimum Deviation Refractometry," *J. Res. Nat. Bur. Std.* **6**, 59 (1931).
126. L. W. Tilton, "Permissible Curvature of Prism Surfaces and Inaccuracy of Collimation in Precise Minimum-Deviation Refractometry," *J. Res. Nat. Bur. Std.* **11**, 25 (1933).
127. L. W. Tilton, "Variations in Refractive Index of CO_2-free Dry Air and a Statistical correlation with Solar Activity," *J. Res. Nat. Bur. Std.* **13**, 111 (1934).
128. L. W. Tilton, "Standard Conditions for Precise Prism Refractometry," *J. Res. Nat. Bur. Std.* **14**, 393 (1935).
129. H. W. Straat and J. W. Forrest, "The Accuracy Requirements in Fifth-Place Refractometry," *J. Opt. Soc. Am.* **29**, 240 (1939).
130. L. W. Tilton, "Testing and Accurate Use of Abbe-Type Refractometers," *J. Opt. Soc. Am.* **32**, 371 (1942).

131. L. W. Tilton, "Sources of Error in Precise Commercial Refractometry," *J. Res. Nat. Bur. Std.* **30**, 311 (1943).
132. L. E. Dodd, "Calibration of Abbe Refractometer with Compensating Prisms, to Measure Refractive Index for Any Wavelength," *Rev. Sci. Inst.* **2**, 466 (1931).
133. J. Guild, "Notes on the Pulfrich Refractometer," *Proc. Phys. Soc.*, (London) **30**, 157 (1918).
134. A. S. Holik, "Refractive Index Measurement of Optically Isotropic Particulate Material by Transmission Nomarski Differential Interference Contrast Microscopy" (Abstract), *J. Opt. Soc. Am.* **72**, 1828 (1982).
135. J. Jelley, "A Microrefractometer and its Use in Chemical Microscopy," *Roy. Mic. Soc.* **54**, 234 (1934).
136. A. C. Candler, *Modern Interferometers*, Hilger and Watts, London, 1951.
137. L. Rayleigh, "On Some Physical Properties of Argon and Helium," in *Progress in Optics*, Vol. 4, *Proc. Roy. Soc. Am.* **29**, 240 (1939).
138. R. W. Wood, *Physical Optics*, Macmillan, New York, 1934.

Spectroscopic Instruments

139. J. F. James and R. S. Sternberg, *The Design of Optical Spectrometers*, Chapman and Hall, London, 1969.
140. P. Bousquet, *Spectroscopy and its Instrumentation*, Adam Hilger, London, 1971. Translated by K. M. Greenland from *Spectroscopie Instrumentale*.
141. A. P. Thorne, *Spectrophysics*, Chapman and Hall, London, 1974.
142. K. I. Tarsov, *The Spectroscope*, (J. H. Dixon, trans.), Hilger, Bristol, 1974.
143. D. Williams, ed., *Methods of Experimental Physics*, Vol. 13, Academic Press, New York, 1976.
144. A. Girard and P. Jacquinot, "Principles of Instrumental Methods in Spectroscopy," in *Advanced Optical Techniques* (A. C. S. Van Heel, ed.), p. 71, Wiley, New York, 1967.
145. E. C. C. Baly, *Spectroscopy*, Longmans, London, 1912.
146. G. L. Clark, ed., *The Encyclopedia of Spectroscopy*, Reinhold, New York, 1960.
147. P. Jacquinot, "Interferometry and Grating Spectroscopy: An Introductory Survey," *Appl. Opt.* **8**, 497 (1969).
148. H. A. Gebbie, "Fourier Transform Versus Grating Spectroscopy," *Appl. Opt.* **8**, 501 (1969).
149. J. E. Steward, "Polarization Interferometer for the Determination of Spectral Modulation Transfer Functions of Monochromators," *Appl. Opt.* **6**, 1523 (1967).
150. K. Kozima, H. Kanamori, and O. Matsuda, "Direct Measurement of Optical Transfer Functions of Spectroscopy System," *Jap. J. Appl. Phys.* **17**, 1271 (1978).
151. H. Fujiwara, "Transfer Function of Spectroscopic System Using A Sinusoidally Modulated Spectrum," *J. Opt. Soc. Am.* **71**, 238 (1981).
152. P. Jacquinot, "The Luminosity of Spectrometers with Prisms, Gratings, or Fabry–Perot Etalons," *J. Opt. Soc. Am.* **44**, 761 (1954).
153. P. Connes, "High Resolution and High Information Fourier Spectroscopy," in *Aspen International Conference on Fourier Spectroscopy* (G. A. Vanasse, A. T. Stair, Jr., D. J. Baker, eds.), AFCRL-71-0019 Special Reports, No. 114, 1970.
154. P. Fellgett, Ph. D. Thesis, "On the theory of infra-red sensitivities and its application to the investigation of stellar radiation in the near infra-red," Cambridge University, Cambridge, 1951.
155. R. E. Poulson, "Test Methods in Spectrophotometry: Stray-Light Determination of Spectral Modulation Transfer Functions of Monochromators," *Appl. Opt.* **6**, 1523 (1967).
156. D. M. Hunten, "Reshaping and Stabilisation of Astronomical Images," in *Methods of Experimental Physics. Vol 12: Astrophysics* (N. Carleton, ed.), p. 193, Academic Press, New York, 1974.

157. R. P. Bauman, *Absorption Spectroscopy*, Wiley, New York, 1962.
158. W. G. Driscoll, "Spectrophotometers," in *Applied Optics and Optical Engineering*, Vol. 5, p. 85 (R. Kingslake, ed.), Academic Press, New York, 1969.

Dispersion Spectrometers

159. J. Strong, "Resolving Power Limitations of Grating and Prism Spectrometers," *J. Opt. Soc. Am.* **39**, 320 (1949).
160. R. A. Sawyer, *Experimental Spectroscopy*, Prentice-Hall, Englewood Cliffs, New Jersey, 1956.
161. G. R. Harrison, R. C. Lord, and J. R. Loofbourow, *Practical Spectroscopy*, Prentice-Hall, Englewood Cliffs, New Jersey, 1948.
162. R. J. Meltzer, "Spectrographs and Monochromators," in *Applied Optics and Optical Engineering*, Vol. 5, Chap. 3 (R. Kingslake, ed.), Academic Press, New York, 1969.
163. G. W. Stroke, "Diffraction Gratings," in *Handbuch der Physik*, Vol. 29, p. 426 (S. Flügge, ed.), Springer-Verlag, Berlin, 1967.
164. A. Girard, "Spectromètre à Grilles," *Appl. Opt.* **2**, 79 (1963).
165. L. Rayleigh, "Investigations in Optics, with Special Reference to the Spectroscope," *Phil. Mag.* Series 5, **8**, 261 (1879).
166. P. Jacquinot and C. Dufour, "Conditions Optiques D'Emploi des Cellules Photo-Electriques dans les Spectrographes et les Interféromètres," *Journal des Recherches du C.N.R.S.* **6**, 91 (1948).
167. K. D. Mielenz, "Spectroscopic Slit Images in Partially Coherent Light," *J. Opt. Soc. Am.* **57**, 66 (1967).
168. M. V. R. K. Murty, "Theory and Principles of Monochromators, Spectrometers and Spectrographs," *Opt. Eng.* **13**, 23 (1974).
169. M. J. E. Golay, "Multi-Slit Spectroscopy," *J. Opt. Soc. Am.* **39**, 437 (1949).
170. M. J. E. Golay, "Static Multislit Spectroscopy and Its Application to the Panoramic Display of Infrared Spectra," *J. Opt. Soc. Am.* **41**, 468 (1951).
171. J. A. Decker, "Hadamard-Transform Spectroscopy," in *Spectrometric Techniques*, Vol. 1, p. 190 (G. A. Vanasse, ed.), Academic Press, New York, 1977.
172. M. Harwit and N. Sloane, *Hadamard Transform Optics*, Academic Press, New York, 1979.

Prism Spectrometers

173. R. Kingslake, ed., "Dispersing Prisms," in *Applied Optics and Optical Engineering*, Vol. 5, Chap. 1, Academic Press, New York, 1969.

Grating Spectrometers

174. G. W. Stroke, "Ruling Testing and Use of Optical Gratings for High Resolution spectroscopy," in *Progress in Optics*, Vol. 2, p. 426 (E. Wolf, ed.), Academic Press, New York, 1974.
175. E. G. Loewen, "Diffraction Gratings," in *Applied Optics and Optical Engineering*, Vol. 9, p. 33, (R. R. Shannon and J. C. Wijaut, eds.), Academic Press, New York, 1969.
176. E. G. Loewen, *Diffraction Grating Handbook*, Bausch and Lomb, Rochester, New York, 1970.
177. D. Richardson, "Diffraction Gratings," in *Applied Optics and Optical Engineering*, Vol. 5, Chap. 2 (R. Kingslake, ed.), Academic Press, New York, 1969.
178. F. Kneubuhl, "Diffraction Grating Spectroscopy," *Appl. Opt.* **8**, 505 (1969).

179. D. J. Schroeder, "Diffraction Grating Instruments," in *Methods of Experimental Physics*, Vol. 12: *Astrophysics*, (N. Carleton, ed.) p. 463, Academic Press, New York, 1974.
180. S. P. Davis, *Diffraction Grating Spectrographs*, Reinhart and Winston, New York, 1970.
181. H. G. Beutler, "The Theory of the Concave Grating," *J. Opt. Soc. Am.* **35**, 311 (1945).
182. W. T. Welford, "Aberration Theory of Gratings and Grating Mountings," in *Progress in Optics*, Vol. 3, p. 241 (E. Wolf, ed.), North-Holland, Pub. Amsterdam, 1965.
183. G. Schmahl and D. Rudolph, "Holographic Diffraction Gratings," in *Progress in Optics*, Vol. 14, p. 193 (E. Wolf, ed.), North-Holland, Publ. Amsterdam, 1976.
184. R. Petit, ed., *Electromagnetic Theory of Gratings*, Springer-Verlag, Berlin, 1970.
185. D. Maystre, "Rigorous Vector Theories of Diffraction Gratings," in *Progress in Optics*, Vol. 21, p. 1 (E. Wolf, ed.), North-Holland, Publ. Amsterdam, 1984.
186. R. W. Wood, *Proc. Phys. Soc.* (London) **18**, 396 (1902).
187. E. G. Loewen, M. Nevière, and D. Maystre, "Grating Efficiency Theory as it Applies to Blazed and Holographic Gratings," *Appl. Opt.* **16**, 2711 (1977).
188. T. M. Hard, "Laser Wavelength Selection and Output Coupling by a Grating," *Appl. Opt.* **9**, 1825 (1970).

Fabry–Perot Spectrometers

189. S. Tolansky, *High Resolution Spectroscopy*, Pitman, New York, 1947.
190. S. Tolansky, *An Introduction to Interferometry*, Longmans, London, 1954.
191. P. Jacquinot, "New Developments in Interference Spectroscopy," in *Reports on Progress in Physics*, Vol. 23 (A. C. Stickland, ed.), The Physical Society, London, 1960.
192. A. H. Cook, *Interference of Electromagnetic Waves*, Clarendon Press, Oxford, 1971.
193. W. H. Steel, *Interferometry*, Cambridge Univ. Press, Cambridge, 1983.
194. F. L. Roesler, "Fabry–Perot Instruments for Astronomy," in *Methods of Experimental Physics*, Vol. 12: Astrophysics (N. Carleton, ed.), Academic Press, New York, 1974.
195. J. R. Greig and J. Cooper, "Rapid Scanning with the Fabry–Perot Etalon," *Appl. Opt.* **7**, 2166 (1968).

Fourier Transform Spectrometers

196. A. Michelson, "On the Application of Interference-Methods to Spectroscopic Measurements," *Phil. Mag.* Ser. 5, **31**, 338 (1891).
197. E. O. Brigham, *The Fast Fourier Transform*, Prentice-Hall, Englewood Cliffs, New Jersey, 1974.
198. L. Mertz, *Transformations in Optics*, Wiley, New York, 1965.
199. R. J. Bell, *Introductory Fourier Transform Spectroscopy*, Academic Press, New York, 1972.
200. J. Chamberlain, *The Principles of Interferometric Spectroscopy*, Wiley, New York, 1979.
201. H. W. Schnopper and R. I. Thompson, "Fourier Spectrometers," in *Methods of Experimental Physics*. Vol. 12: *Astrophysics* (N. Carleton, ed.), p. 491, Academic Press, New York, 1974.
202. G. A. Vanasse and H. Sakai, "Fourier Spectroscopy," in *Progress in Optics*, Vol. 7, Chap. 1 (E. Wolf ed.), North-Holland Publ., New York, 1967.
203. H. Sakai, "High Resolving Power Fourier Spectroscopy," in *Spectrometric Techniques*, Vol. 1, Chap. 1 (G. A. Vanasse, ed.), Academic Press, New York, 1977.
204. R. J. Bell, "Applications of Fourier Spectroscopy," in *Spectrometric Techniques*, Vol. 1, (G. A. Vanasse, ed.), Academic Press, New York, 1977.

205. J. B. Breckinridge and R. A. Schindler, "First-Order Optical Design for Fourier Spectrometers," in *Spectrometric Techniques*, Vol. 2, p. 63 (G. A. Vanasse, ed.), Academic Press, New York, 1981.
206. G. A. Vanasse, A. T. Stair, Jr., D. J. Baker, Ed., Aspen International Conference on Fourier Spectroscopy, 1970, AFCRL-71-0019 Special Reports, No. 114.
207. J. Connes, "Spectroscopic Studies Using Fourier Transformations," NOTS Technical Publication 3157, translated from "Recherches sur la Spectroscopie par Transformation de Fourier," *Rev. D' Optique* **40**, 45, 116, 171, 231 (1961).
208. D. J. Lovell, "Performance and characteristics of Commercially Available Instruments," in *Spectrometric Techniques*, Vol. 1 (G. A. Vanasse, ed.), Academic Press, New York, 1977.
209. P. Jacquinot and B. Roizen-Dossier, "Apodisation," in *Progress in Optics*, Vol. 3, Chap. 2 (E. Wolf, ed.), North-Holland Publ., Amsterdam, 1964.
210. R. Bracewell, *The Fourier Transform and its Applications*, McGraw-Hill, New York, 1978.
211. G. Guelachvili, "Distortions in Fourier Spectra and Diagnosis," in *Spectrometric Techniques*, Vol. 2, p. 1 (G. A. Vanasse, ed.), Academic Press, New York, 1981.
212. A. S. Zachor, I. Coleman, and W. G. Mankin, "Effects of Drive Nonlinearities in Fourier Spectroscopy," in *Spectroscopic Techniques*, Vol. 2, p. 127 (G. A. Vanasse, ed.), Academic Press, New York, 1981.
213. D. Baker, "Field-Widened Interferometers for Fourier Spectroscopy," in *Spectrometric Techniques*, Vol. 1 (G. A. Vanasse, ed.), Academic Press, New York, 1977.
214. P. Bouchareine and P. Connes, "Interféromètre à Champlomepense pour Spectroscopie par Transformation de Fourier," *J. Phys. Radium* **24**, 134 (1963).

5. LIGHT SOURCES

Zacarias Malacara H.
and
Arquimedes A. Morales R.

Centro de Investigaciones en Optica A. C.
Apdo. Postal 948
37000 Leon, Gto. Mexico.

5.1. Introduction

Light sources produce radiation, known as light, which most of the time may be detected visually. Some light sources have spectral characteristics producing radiation that cannot be seen. Therefore, even though some of the sources emit in the ultraviolet (UV) and infrared (IR) portions of the spectrum, they are included here. A complete description of the subject can be found in Elenbaas[1] and the IES Handbook.[2] A historical account is described by Anderson and Saby.[3]

The sun is the most important source of light, used in instrumentation since the ancient Egyptian ages. Its spectral distribution is very wide, with some absorption bands which are due mainly to atmospheric absorption.[4,5] For its importance, the sun deserves a full treatment, which has been done in many books.[6,7]

5.2. Filament Sources

5.2.1. Tungsten Filament Lamps

The filament light source is in a continuous stage of development, and new improvements are released every day. The popularity of filament lamps lies in their low cost, availability, simplicity of operation, and reliability.

An incandescent metal wire inside a glass envelope, filled with an inert gas, continues to be the basic lamp for optical applications. Some wires are used both as holders and as electrical conductors to a base, which support the lamp in position and serve as electrical port.

The components of an incandescent lamp which work any time the lamp is turned on are:

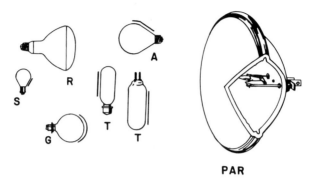

FIG. 1. Commercial incandescent lamp shapes. (Adapted from Ref. 19.)

1. *Bulb.* The bulb, originally designed to maintain a vacuum, and later, to enclose an inert gas, has evolved into an active optical component in instrumentation. The shape of the bulb ranges from the arbitrary one (A) to the reflector (R), the tubular (T), or the parabolic (PAR). Some common glass bulbs are shown in Fig. 1.

The bulb is made with a frosted or a transparent surface. In the PAR and R shapes, the bulb is reflectorized. In recent versions, a thin-film coated parabolic reflector (cold mirror) is attached to the bulb. In such lamps, a heat decrease of 50% is achieved at the specimen or film gate. Common bulb materials are crown glass and, in some cases, fused silica. Optical transmittance curves have been published elsewhere.[8,9]

2. *Filaments.* Tungsten is the most common material used for filament wires, where the dissipated electrical power increases the temperature on the filament to about 3700°K, making it to radiate as a gray body with an emissivity between 0.42 and 0.48. Spectral dependence on tungsten emissivity is reported by Larrabee[10] and Forsythe and Adams[11] for different temperatures. Figure 2 shows these results. Some of the common filament shapes are shown in Fig. 3. Filaments approximate a Lambertian source. This is especially true for ribbon filaments, but for coiled filaments, there is a departure from this approximation.

Coiling the filament increases its light-producing efficiency by about 10%. Double coiling further increases the efficiency. Irradiance distribution at some distance from the filament is determined by the filament form and other parameters, like the lamp envelope, with the base having a secondary effect.[4,12] The intensity in a particular direction coming from an incandescent source, such as a tungsten filament, is proportional to the area of filament that can be seen in that direction and to its luminance.[13] In all planes that include the straight-wire filament axis, the intensity varies approximately following Lambert's law.

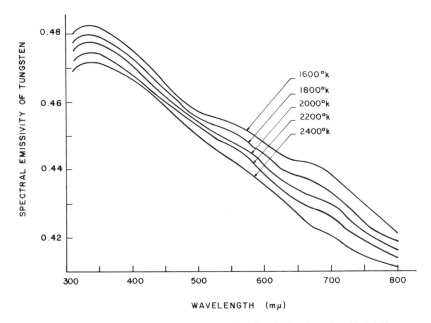

FIG. 2. Spectral dependence on tungsten emissivity. (After Larrabee, Ref. 10.)

Heat is dissipated from the filament by the inert gas convective currents inside the bulb, causing it to depart from a blackbody. For coiled filaments, emissivity is closer to a blackbody.

Electrical resistance increases with temperature. The largest power is dissipated during turning on of the bulb, making the failure rate higher in systems that switch from cold very often. This can be reduced by limiting the current in the filament during the startup or by maintaining a small current during off-time. The lamp fails when the evaporated tungsten reduces the filament thickness to such an extent that the increase in temperature fuses the metal. In high-vibration environments, the filament tends to break because of the metal softening. This could be avoided by using lamps with a large number of holders, or it could be minimized by installing the lamp with the filament along the vibration axis.

3. *Filling gas.* Usually an inert gas is used to fill the lamps, commonly a mixture of nitrogen and argon. It reduces tungsten evaporation and allows higher-temperature operation for the filament.

4. *Lead-in wires.* They make the electrical connection of the tungsten filament to the lamp base. The thermal expansion coefficient must be equal to that of the glass in the stem press. This can be achieved by a nickel iron alloy core in a copper sleeve.

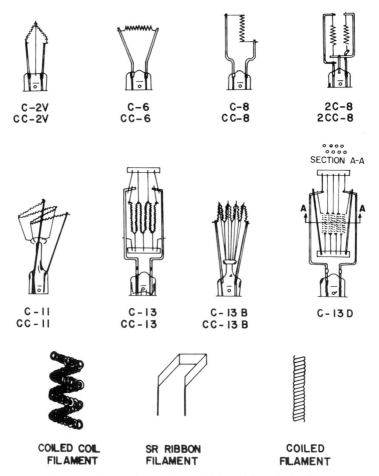

FIG. 3. Some filament shapes. (Adapted from Ref. 19.)

5. *Support wires.* They hold the filament in place and protect it during shipment and service. These wires are made of molybdenum, and their use is not recommended because they reduce lamp efficiency. However, for lamps in vibrating environments, several support wires must be used.

6. *Base.* The base in a lamp is used both as an electrical contact and as a support for the lamp itself. The simplest base is of the screw type, like the candelabra and mogul bases. Other bases have the purpose of positioning the filament with respect to an optical axis, as in the so-called prefocus base. The latter is preferred in optical instruments. Examples for the most commonly used bases are shown in Fig. 4, taken from Carlson and Clark.[4]

FILAMENT SOURCES 243

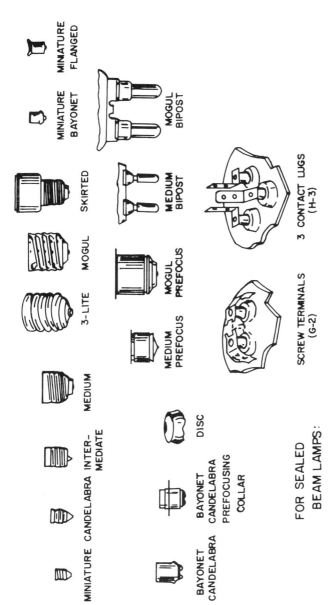

FIG. 4. Some bases for incandescent lamps. (Adapted from Ref. 4.)

Several factors, are worth taking into account in a lamp. The filament evaporates at a nonconstant rate, which depends on the amount of impurities in the material as found by Larrabee.[10] As an immediate consequence, such material is deposited in the upper surface on the bulb. In optical applications, the lamp must be operated vertically with the light beam exiting through the sidewall. In some lamps, a screen is included to collect the evaporated material, allowing the operation in positions other than a vertical one. The tungsten evaporation reduces the lamp resistivity, reducing also the radiance A more constant radiance can be achieved by supplying a constant current to the lamp, instead of a constant voltage. In this case, the radiance is maintained at a steadier rate, but the lamp life is considerably reduced. When changing the nominal voltage, many parameters, like the resistance, current, power, and lumens-per-watt also change, as shown in Fig. 5 (after Carlson and Clark[4]). The intensity of a filament source can be modulated, obtaining a frequency variation of the intensity equal to twice the driving-current frequency (assuming no direct current components on the modulating current). The amplitude of the modulation as a function of the frequency has been reported by Tolles and Ihring.[14] It should be noticed that the modulation is a function of the wavelength because of the change in the color temperature as the filament luminance extinguishes.

Filament design is a balance between light output and life. Lamp efficacy is almost directly related to the filament temperature. The most efficient lamps have the shortest lives, and those with extremely long lives are the least efficient.

In optical instrumentation, like spectroscopy and retinoscopy, it is often desirable to have a line-shaped source. A carefully chosen line filament is then selected instead of using pinholes or slits and lenses. Some lamps have special straight and short filaments, for example those used in retinoscopes (Welch-Allyn No. 03700). These lamps can be very useful in optical instruments, where a bright linear light source is desired. In other applications, like microscopy and projection systems, an extended source is required to fill the specimen or the film gate. In such cases, double-coiled filaments are used [Fig. 3]. Optical requirements for those optical systems are described by Kingslake.[15]

Some common lamps used in optical instrumentation have a conic reflector behind them, as shown in Fig. 6. This reflector has a large collecting solid angle and focuses the light at some point in front of the lamp. Some lamps have the focal point at 5.6 inches from the rim of the reflector. These lamps have a corrugated reflector surface and a filament (CC-8) coiled along the axis in order to produce an extended and diffuse spot of light. An example of application of these lamps are on the slide projectors. A second type of lamp with a conic reflector has the focal point at distances

FILAMENT SOURCES

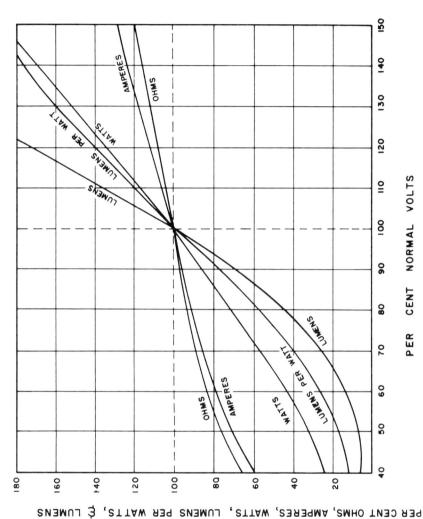

FIG. 5. Dependence of the resistance, current, power, and lumens per watt on the input voltage.

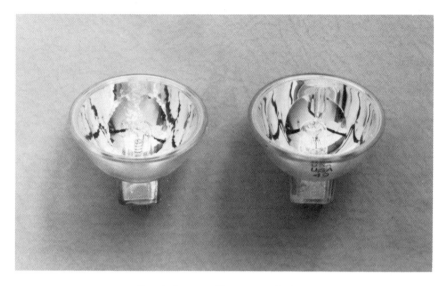

FIG. 6. Lamps with a conic reflector.

between 1.1 and 5.5 inches from the rim of the reflector. The surface of the reflector in these lamps is smooth and has a small filament (CC-6) in order to produce a relatively small spot of light. The most common use of these lamps is in optical fiber illuminators.

5.2.2. Tungsten Halogen Lamps

When a high-pressure halogen vapor, typically iodine, is added to the inert gas, the halogen gas reacts with the vaporized tungsten. The high filament temperature reverses the reaction, causing the tungsten to redeposit on the filament, freeing halogen vapor. This is the so-called halogen cycle, which extends the lamp life and avoids bulb blackening due to evaporated tungsten. To keep the halogen cycle, high temperature is needed even in the bulb surface. These lamps are made small to keep a high temperature in the whole lamp, making it necessary to use a quartz bulb to support high temperatures. Tungsten halogen lamps require at least 250°C to operate. Due to the high temperatures involved, skin grease or other impurities in the bulb surface may produce devitrification on the quartz, which can be avoided by cleaning the cold surface with a solvent. In any case, touching the surface directly with the fingers should be avoided. In halogen lamps, molybdenum leads are used, which react with the quartz, producing oxidation of these leads. This is the main cause of failure in the quartz–halogen lamps. The oxidating reaction increases with temperature and is cumulative with time. Halogen vapor has a higher pressure than the atmosphere. It has been found by Studer and VanBeers[16] that halogen vapor absorbs some radiation in the yellow-green portion of the spectrum, and that it emits in the blue and ultraviolet.

The stability of the halogen lamps has been found convenient as a secondary standard for irradiance, with a maximum stability of 3% for the visible spectrum. Calibration for the tungsten-halogen lamps is reported by Stair et al.[17] Commercially available secondary irradiance standards are available using a so called type FEL lamp and are described in a National Bureau of Standards (NBS) publication, as mentioned by Grum and Becherer.[18]

5.2.3. Photoflash Lamps

These are special versions of lamps designed to produce very high levels of light at about 3400°K to provide color-corrected light for photographic film (General Electric Co. catalog[19]). The old disposable lamps were built with a long aluminum or magnesium filament in a bulb filled with oxygen. The hot filament ignites a primer material producing a fast combustion with high luminances. Sometimes capacitors had to be used for a fast energy supply to the lamp. Newer designs enclose several lamps in a single package, where the battery has been replaced by a piezoelectric element that fires the primer. In other lamps, the primer is fired mechanically from the camera. A brief description of these lamps are found in Carlson and Clark[9] and in Eby and Levin.[12]

5.3. Arc and Discharge Sources

Light can also be produced by electron de-excitation in gases. Gas excitation can be obtained by electron or ion collision, which in turn, is achieved by some of the following means:

(a) Thermal electron emission by a hot cathode: By heating an incandescent filament, some of the conduction electrons get a higher energy than the metal's work function and are thus released. A high potential accelerates the electrons, and by collision, new ions are produced. This method is used mainly to start a discharge.

(b) Secondary emission: Electrons are released by the collision of positive ions on the cathode. Once the discharge is initiated, it is sustained by the continuous ion impacts. This mechanism is developed in arc sources.

(c) Field emission: A highly electrical field is applied to an initially cold electrode. Once the current is started, it is maintained by secondary emission. This process is used to start a discharge in arcs with an auxiliary starting electrode.

A cold arc lamp has a high impedance, but once ignition starts, the device exhibits very low impedance. Then, these lamps must be operated with a ballast or other current-limiting device. Inductive ballasts are appropriate for ac supplies. For dc, a resistive ballast or current regulation or both must be used. Electrical dynamical characteristics for discharge lamps are discussed by Francis.[20] A method for optical stabilization by current regulation is suggested by Pulfrey.[21]

5.3.1. Carbon Arc Sources

Carbon sources have been used successfully for many years for high-intensity illumination, searchlights, and movie projection. They get their name from the bow-shaped discharge between the carbon electrodes. Under discharge, electron and ion bombardment takes place in both electrodes, the brighter one being the positive crater. Electrodes are consumed at a rate that prevents its use for long time periods; hence, a mechanical system has been constructed by Latil[22] to extend its operational time.

Arcs are classified according to their discharge of either high intensity, low intensity, and flame.[4] Most of the carbon arcs are operated from a 120 V supply. The low- and high-intensity arcs are operated on dc, but the flame type can be operated on both ac and dc.

The main source of radiation is the anode tip for the low-intensity arc, the positive crater and surrounding gas for the high-intensity arc, and the flame for the flame arc.

Electrodes are carbon rods, but some impurities are added to change the spectral distribution. Carbon arcs are essentially blackbody emitters with an emissivity of about 0.98 to 0.99 (Null and Lozier[23]), and the color temperature varies from 3800 to 6500°K. Spectral distribution in an arc source is very close to a gray body except for some lines in the ultraviolet, as measured by Null and Lozier.[23] Care must be taken for protecting the eyes from UV when using this source.

The main disadvantage of the carbon arc is the high consumption rate that requires the separation of the electrodes to be carefully controlled, and the need to operate with bulky ballasts. It has been suggested by Null and Lozier[23] that a properly operated carbon arc can be used as a radiation standard.

5.3.2. Compact Source Arc

The high consumption rate and relative unstability of the carbon arc led to the development of what is known as the *compact source arc*. A description for it follows.

A spherical glass bulb encloses an inert gas, usually at atmospheric pressure, but when hot, pressure may increase up to 50 atmospheres. Two tungsten or copper electrodes, separated by a distance from 1 mm to 1 cm, produce between them the desired arc. Sometimes, the bulb is enclosed inside another larger bulb, as shown in Fig. 7.

Some care has to be taken when using these lamps, because of the high pressures involved. Appropriate housing must also be used, and windows must be installed along the optical path for protection for UV. Ventilation is needed to dissipate produced ozone, although some lamps are made with UV-filtering glass.

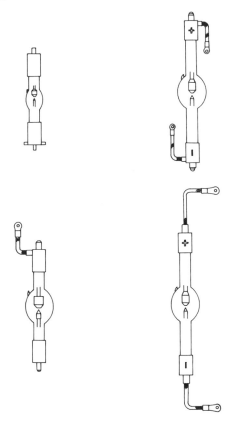

FIG. 7. Typical compact source arc.

Compact source arcs can be operated from ac to dc sources. In dc lamps, the anode is usually made larger to dissipate the higher temperature on this electrode; in ac lamps, both electrodes are the same size. High voltage of the order of 40 kv is provided to start ignition. Mercury and mercury-xenon lamps take about 15 minutes to vaporize mercury, and once vaporized, pressure increases to about 50 atmospheres. Restarting of the lamp can only be done after cooling of the lamp in order to reduce the gas pressure. In some lamps, an additional electrode is included for starting. Lamp life is rated at more than 1000 hours. Life ends generally with bulb blackening from evaporated metal. When power is added to compensate for the light decrease by aging, color temperature decreases at a rate of 200°K after 1000 hours, according to Eby and Levin.[12] Under such conditions, violent failure may occur.

Because the high pressures involved in these lamps, lines tend to broaden and a continuous background appears. Hot electrodes contribute some red coloration. A typical luminance distribution in the arc is shown in Fig. 8, reproduced from the Hanovia catalog.[24] Radiance considerations are found in Carlson and Clark.[4]

Three gas mixtures are used to fill the bulb: mercury, mercury-xenon, and xenon. Compact source lamps are commercially available from 75 to 30,000 watts. Sources with more than 5000 watts usually have a water-cooled anode.

Mercury lamps emit the characteristic lines from mercury and some continuous background, with most of the radiation in the ultraviolet. Color temperature is about 8000°K.

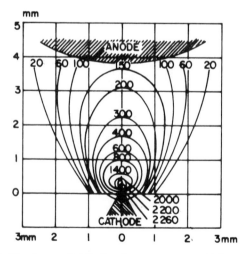

FIG. 8. Spectral luminance distribution for an arc source. (Adapted from Ref. 24.)

ARC AND DISCHARGE SOURCES 251

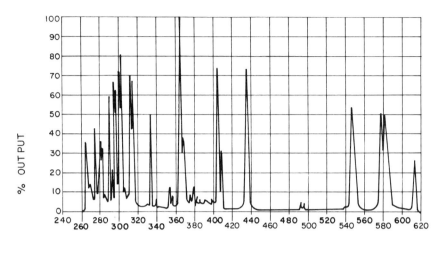

FIG. 9. Spectral emission for a mercury-xenon lamp. (Adapted from Ref. 24.)

Mercury-xenon lamps emit about 35% of their energy in the ultraviolet. Xenon is used for starting and maintaining the arc, while mercury evaporates. If restarting is tried when the lamp is still hot, the igniting spark must have a very high voltage because of the high pressure. Light efficacies for mercury-xenon lamps are about 55 lumens per watt. The spectral emission curve for mercury-xenon lamps is shown in Fig. 9.

Xenon lamps emit an essentially continuous spectrum in the visible portion, with some lines in the near infrared, as shown in Fig. 10.[24] Since no metals are vaporized, full output is reached shortly after the start. Color temperature is about 6000°K, and light efficacies are about 35 lumens per watt. With a high enough voltage, this lamp can be restarted while still hot. Stability after lamp aging for xenon lamps has been found to be within 0.1% by Budde.[25]

5.3.3. Discharge Lamps

In a discharge lamp, a current flows in a low-pressure gas. Some examples for these lamps are fluorescent lamps, spectral lamps, and flash tubes.

Fluorescent lamps find few applications in optical instrumentation, their main use being for lighting. An extensive description is given by Pritchard,[26] and Frier and Frier.[27] These lamps are usually made in a glass tube of 152-mm to 2438-mm length. Emitted light has a mercury spectrum, the strongest line being the mercury line at 0.2537 µm. To increase light

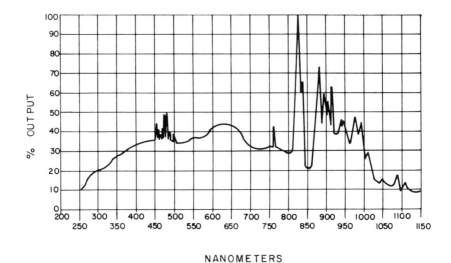

FIG. 10. Spectral distribution for a xenon lamp. (Adapted from Ref. 24.)

efficiency, a phosphor mixture is used to reradiate the 0.2537 μm line into the visible, obtaining light efficacies as great as 83 lumens per watt. Several phosphor mixtures are available to obtain different color temperatures. Spectral distribution for some phosphorus is found in the *IES Lighting Handbook*[2] and in Grum and Becherer.[18] Fluorescent lamps exhibit modulation from line supply. Since color temperature is a function of voltage, modulation changes with wavelength.

Basically, there are three types of lamps, depending on the way the lamp is started: (*a*) preheated, (*b*) instant start, and (*c*) rapid start. In a preheated lamp, the sockets have two pins connected to a tungsten filament. When power is first applied to the circuit, a starter switch is closed and the lamp filaments are connected in series with the ballast in order to preheat the lamp cathodes or filaments. After a short time, the starter switch is opened, causing a transient voltage high enough to start the lamp.

In the instant-start lamp, the sockets have only one thick pin. In these lamps, the ballast contains a transformer that provides sufficient open-circuit voltage to attract electrons from the cathodes without preheating.

The rapid-start lamps have sockets with two pins. They have separate windings to heat the lamp cathodes continuously during the lamp's operation. The heating of the electrodes reduces the voltage necessary to strike the arc and improves the efficiency of the lamp.

Some specially designed fluorescent lamps are: the black-light lamp, in which a phosphor is included to reemit the 0.2537 μm line to 0.3650 μm in the near ultraviolet. The glass bulb is designed to transmit this wavelength. Another type is the germicidal lamp, which does not have any phosphor, and the glass bulb is designed to transmit UV. This lamp is frequently used to expose photoresists, to eliminate germs, and to erase EPROM (Eraseable Programmable, Read Only, Memory).

An interesting type of fluorescent lamp is found in photocopiers, where a reflective coating is included in a region of the tube to increase the luminance along a defined axis. Two approaches are shown in Fig. 11, as described by Eby and Levin.[12]

Spectral lamps have been widely used by spectroscopists. Inert gases and some metals are included in a discharge source. They emit discrete lines, and low pressure is maintained to keep the lines narrow and hence a high light temporal coherence. It has a typical value in the case of some mercury lines of 0.3 to 3 cm as pointed out by Welford,[28] depending on the pressure and temperature of the lamp. These lamps can be used as wavelength standards, as was suggested by Michelson and Morley.[29] On trying to find the best source, Michelson discovered the hyperfine structure in mercury, as described by Meggers.[30] For this reason, it was decided to abandon the mercury lines as length standards and to adopt a cadmium line, which was later to be substituted by a Krypton line. For specific characteristics of each spectral lamp the manufacturers' technical data should be consulted.[31,32,5]

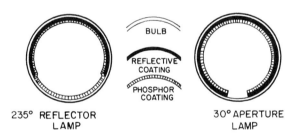

FIG. 11. Transverse cross-sectional view for some directional fluorescent lamp. (From Ref. 12.)

Flash-discharge lamps usually have a straight, U-shaped or coiled glass tube. Electrodes have a spacing of several centimeters, requiring large voltages to flash. Stored electrical energy in a capacitor is released in a very short time (several nanoseconds). These lamps, filled with xenon, are frequently found in photographic lamps, with a color temperature of 6000°K, Xenon flash lamps are also used for laser pumping in ruby, neodymium, and YAG (Ytrium Aluminum Garnet) lasers. In laser applications, both coiled and straight tubes are used. In some high-power sources, water-cooled jackets remove excess heat. For large lamps, a trigger wire is wrapped around the tube. Most of the flash lamps for laser application show a continuous spectrum that approaches that of a blackbody. Pulse repetition rates in lasers are still limited by excess power removal from the lamps and the active element. General lamp requirements for laser applications are discussed by Lengyel,[33] and theoretical calculations for lamp design are published in several papers.[34-36] while practical aspects are found in several sources.[37-40]

5.4. Light-Emitting Diodes

The light-emitting diode (LED), is a solid state lamp, made of a p-n semiconductor junction, usually gallium arsenide and dopants. The actual construction depends on the desired emitted wavelength, luminous power, emission characteristics, and life.

Charge recombination takes place in a semiconductor junction, with the emitted wavelength being a function of the semiconductor's band gap. Details about the construction and the physics of light production is left to other reference.[41-44] Under forward bias, an LED conducts current as any other p-n junction device and radiates in the infrared or visible spectral region.

Table I shows the wavelengths and bandwidths for commercial LEDs.[41] Difficulties have arisen with the mass production of blue-emitting LEDs, although production on a laboratory scale is reported by Kawabata et al.[45] The luminous intensity of an LED is nearly linear with the current. Actual dependence of the luminous intensity on the current for a typical device is shown in Fig. 12.

Commercially, an LED is available with two different-color LEDs in the same package, so that the color changes from red to green by changing the direction of the current. When red and green are mixed by using an oscillating current source, the apparent color is yellow or orange, with the color depending on the duty cycle of the waveform. An extension of this combination of different emitters led to the development of other bipolar

TABLE I. Some Characteristics of LEDs

Material	Color	Wavelength nm	Bandwidth at 1/2 peak	Rise time sec	Luminance cd/m^2
GaAa:Si	IR	940		$3 \times 10E-7$	
GaAs	IR	900	40	$10E-9$	
GaP	Red	690	90	$5 \times 10E-8$	
GaAs$_6$P$_4$	Red	650.67	40	$10E-9$	2569
Ga$_5$As$_5$P	Orange	610	40	$10E-9$	685
SiC	Yellow	590	120	$10E-7$	137
GaP	Green	560	40	$5 \times 10E-8$	1028
GaAs:Si+P	Green:IR	540 & 940	5 & 80	0.004 & $10E-4$	137

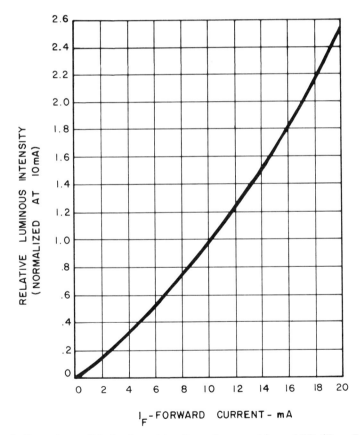

FIG. 12. Dependence of the luminous intensity on the current for an LED. (From Ref. 42.)

devices red/red, green/green, etc. These actually consist of two diodes in each package, allowing either ac/dc or tri-state indication. Typical luminance for commercial LEDs, ranging from 0.5 mcd for the green to 70.0 mcd for the yellow, is obtained from selector guides of some manufacturers.[41,42,43] LED chips are Lambertian emitters, but this can be a disadvantage for some applications. In display lamps, diffuser plastic encapsulation is used to increase the radiation angle, whereas in fiber optics, a directional radiating pattern is desirable. For such cases, the Burrus diode[43] or an edge-emitting diode is used. Typical radiation patterns for commercial lamps for display purposes are shown in Fig. 13.

Several lamp packages have been developed by the optoelectronics industry, with a wide range that covers most common applications. Common

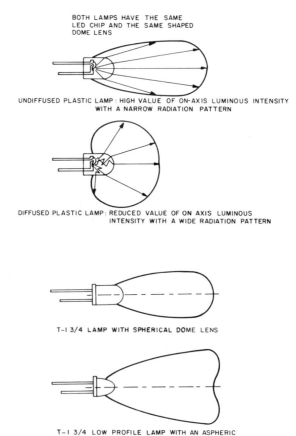

FIG. 13. Typical radiation patterns for commercial LEDs. (From Ref. 42.)

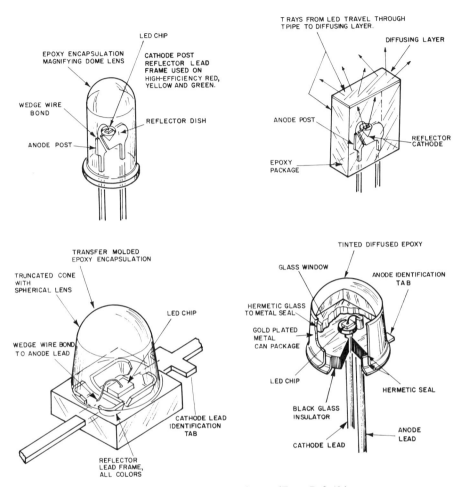

FIG. 14. Commercial LED shapes. (From Ref. 42.)

sizes and shapes are shown in Fig. 14. The industry standard T-1, T-3/4, and T-1 3/4 LEDs are in widespread used. In this designation, the number indicates the lamp diameter in 1/8th of an inch. Other sizes and shapes are available for which the appropriate manufacturer's catalog should be consulted. Light efficacy for a given package is calculated by Jarominsky.[46]

Electrically, an LED lamp has a very small dynamic resistance above the turn-on voltage and must be driven by a current source. For most applications, a resistor can be connected in series with the lamp or another active current source. In some lamps, a resistor for a given voltage is included in the LED package.

Light output is increased by driving the LED with current pulses by increasing the current, and by reducing the duty cycle. This effect is due to the high peak current efficiency for the same average power and the light enhancement factor. The light enhancement factor[41] is attributable to the eye's retention of high-speed and high-brightness pulses. In pulsed operation, a delay of about 200-300 psec is reported by Holden *et al.*[47] between the electrical pulse and the optical signal. A method for contrast-enhancement calculation is described by Pucilowsky *et al.*[48]

The largest usage of LED lamps is in visual display applications. LED lamps have long been used as panel-mounted indicators, printed-circuit status indicators, and both x/y-addressable and linear arrays. With the introduction of high-efficiency LED lamps, applications that previously could only use neon and incandescent lamps, now can also use LED lamps. These high-intensity applications include backlighting a legend or illuminating a push button. LED lamps offer many advantages to the designer. They are small, light weight, and mechanically rugged. LEDs have been proposed by Dawson[49] for fiber optics communications because of their low pulse widening. Some care must be taken in visual applications, where color anomalous observers are present. Such cases are reported by Allyn *et al.*[50]

5.5. Laser Sources

For a long time, physicists have looked for a highly monochromatic optical source or a point source. The appearance of the laser simultaneously solved both of these requirements to a great extent. The most remarkable properties of the laser light are its monochromaticity and source size, both related to the temporal and spatial characteristics, respectively.

5.5.1. Temporal Characteristics

High spectral purity is often required in applications such as interferometry, holography, or optical mixing. For such applications, lasers are the most adequate instruments.

In interferometry and holography, it is suitable to discuss the bandwidth properties of a light source by defining the "coherence length." This concept refers to the maximum usable optical path difference. If a laser is observed in a Michelson interferometer, interference fringes disappear at a certain optical path difference. This optical path difference occurs when its magnitude is equal to the length of the laser cavity. The minimum of fringe visibility with path difference occurs at integral multiples of the laser cavity length.

The output from a laser is usually composed of several discrete frequency components, each component having a very high degree of monochromaticity. The ultimate in spectral purity is achieved when the laser is operated at a single-frequency output. In a given laser system, this requires careful tuning of the optical cavity to the frequency and bandwidth of the amplifying transition. After single-frequency operation is obtained, an increase in spectral purity may be obtained by carefully controlling the mechanical and thermal stability of the optical cavity.

In more critical applications, such as long-path-length interferometry or optical heterodyning, it is usually necessary to have single-frequency operation of the laser.

We have asserted that the spectral and spatial characteristics of lasers are dominated by the properties of the optical cavity. The explanation is made using the concept of modes. A *mode* of an electromagnetic field in a laser cavity is a stationary configuration which is satisfied by the boundary conditions imposed on the mirrors. Field configurations which do not satisfy these boundary conditions are not physically allowed (see Fig. 15). Mode fields, on the other hand, exist within the cavity and interact with the amplifying medium in the cavity.

Because of the relationship between the resonator dimensions and the wavelength, an optical cavity is inevitably excited on a very high overtone. Many of the overtone resonances typically fall within the bandwidth of a laser amplifier. By choosing a structure with open sides, the losses for most of these overtone resonances can be made so high, that oscillation will be possible for only a few modes. Nevertheless, gas lasers employing a single pair of mirrors for feedback are multimode devices. In these devices, the output exhibits several frequency components.

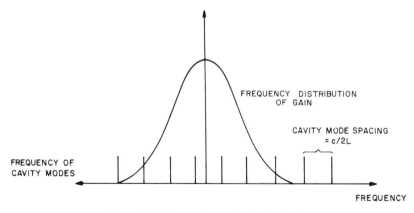

FIG. 15. Cavity modes and gain distribution.

To restrict laser oscillations to a single frequency, additional boundary conditions on the mode structure are usually fulfilled by employing sharp spectral filters such as interference filters. Sometimes it is possible to use a multiple mirror setup which resembles the typical familiar optical interferometers, or a combination of interferometers. It is always better to control the laser frequency spectrum from within the laser cavity. With trying to filter it externally arises the difficulty of matching the center of the passband to a particular mode frequency. Higher single-frequency power results if the laser amplifier medium interacts with a single-cavity mode only.

In contrast to the multimode laser where the coherence length is effectively the length of the cavity, coherence lengths of tens of meters are readily available from single-frequency lasers of reasonable dimensions. With a special effort concerning the dimensional stability of the cavity, the coherence length may be increased to tens of kilometers. Gas lasers have been employed as optical frequency standards, and the developmental work in this area continues.

5.5.2. Spatial Characteristics

The divergence of the laser beam is expressed in terms of the full angle beam divergence, which is twice the angle that the beam's outer edge forms with the center of the beam. The divergence tells us how rapidly the beam spreads out when it is emitted from the laser. The divergence angle is specified in radians. For a typical small laser, the beam divergence is about one milliradian (mrad), and it can be shown that this typical laser beam increases in size by one millimeter for every meter of beam travel. The divergence of a laser beam follows from the intrinsic size of the beam within the laser's optical cavity. One can predict how the beam spreads or changes as it propagates by using the Huygens principle. In most lasers, the beam is circularly symmetric in cross section, being most intense at the center and falling off as $\exp[-(r/w)^2]$ with the distance r from the center of the beam. Such a beam is called a *Gaussian beam*. The value of r, for which the beam irradiance (the power per unit area) decreases to $1/e$ of its value at the center, is termed the spot size w. In general, w will vary from point to point along the axis of the beam. At one point in the cavity, called the *beam waist*, the Gaussian beam has its minimum spot size, w_0, as shown in Fig. 16. The spot size at the beam waist determines the divergence of the laser beam.

The transverse structure of a mode is found to be the same in all planes normal to the optical axis of the cavity, apart from an overall change of scale. Since the output of a laser consists of one of the traveling-wave components of the mode field, we may readily appreciate that the charac-

LASER SOURCES 261

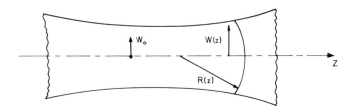

FIG. 16. Beam waist.

teristic structure of a laser beam will remain unaltered, apart from a change in scale, when processed by an ordinary optical system (provided the system aperture is large compared with the transverse extent of the beam). In most applications, it is desirable for this structure to be as simple and uniform as possible. Therefore, it is normal practice in gas-laser design to arrange for operation of the lowest order, the TEM_{00} mode. In fact, it can be shown that while a given laser system and cavity will deliver more total power into higher-order modes, when the output from such a laser is focused by a lens, the illumination in the focal plane (or any other plane) is independent of the order number of the mode.

Since the smallest transverse spatial extent in a given cavity is the TEM_{00} (Transverse Electric Mode), it is normal practice to reduce the aperture of the laser cavity to increase the losses for high-order modes and to favor oscillation in the TEM_{00} mode. Often the laser amplifier geometry will accomplish this directly with the appropriate cavity design.

If we take the direction of propagation to be aligned with the z axis, then the variation of the spot size is described by

$$w^2(z) = w_0^2 \left[1 + \left(\frac{\lambda z}{w_0^2} \right)^2 \right], \quad (5.1)$$

taking the origin of coordinates to be at the beam waist (Fig. 15). This expression applies to the mode field inside the laser cavity as well as to the external field. Of course, each time the beam passes into an optical system with focusing properties, one must find the new location of the beam waist before applying Eq. (5.1) to the system. We will show how to do this shortly (Svetlik,[51] Li and Wolf[52]).

Equation (5.1) describes one aspect of the divergence of a gas laser beam. With reference to the coordinate system established above, the radius of curvature of a wave front at some point z is

$$R(z) = z \left[1 + \left(\frac{w_0^2}{\lambda_z} \right)^2 \right]. \quad (5.2)$$

5.5.3. Other Properties of Lasers

In the strict sense, the term intensity applies only to the amount of power radiated per unit solid angle in watts/steradian (radiometric), or lumens/steradian (photometric). The quantity that describes the power incident on a unit area is called the irradiance in radiometry (watts/m^2) and the illuminance in photometry (lumens/m^2). For example concerning laser power measurements, see O'Shea et al.[53]

5.5.4. Gas Lasers

The operation of these laser devices is carried out by certain excitation. Although the bandwidth of a single laser mode can be as narrow as a fraction of a hertz, the actual oscillation bandwidth of a laser, with no mode control will be governed by the width of the gain curve, which may be several gigahertz for a gas laser in the visible region of the spectrum and much greater for solid-state lasers.

Some frequency selection techniques can be used to confine the laser oscillation to a single atomic or molecular line, as described in the following.

(a) A laser mirror formed by a multilayer dielectric structure is normally designed to be highly reflecting at the laser wavelength. It is also possible to adjust the multilayers so that the mirrors are highly transmitting at the unwanted wavelengths. Then, a low Q is obtained for the undesired wavelengths, and no spurious oscillation will build up.

(b) A prism is used inside the laser resonator with a coated side for frequency selection.

(c) A diffraction grating can also be used.

(d) A gas cell is used in the resonator cavity, which is transparent for the desired laser wavelength but absorbing at the unwanted laser transition.

(e) A Lyot filter inside the resonator cavity is used to obtain a narrow spectral response. High transmission occurs only for those wavelengths for which the rotation of the polarized light is transmitted with low loss.

(f) Phase grating in a solid laser medium can be used to form a "distributed feedback" resonator which will have low losses only over a narrow band of frequencies, as described by Smith.[54]

For a schematic description of these rough frequency selection methods, see Fig. 17. The characteristics of some pulsed and continuum gas lasers are shown in Tables II and III.

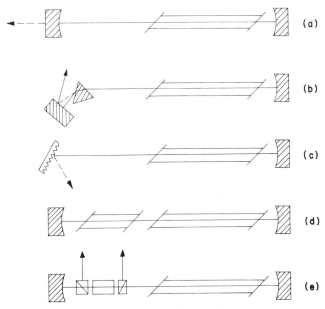

FIG. 17. Coarse frequency selection methods: (a) multilayer mirrors, (b) prism inside the cavity, (c) diffraction grating, (d) gas cell in the cavity, (e) Lyot filter.

5.5.5. Dye Lasers

Such lasers are very attractive because they can be easily tunable over a wide range of frequencies. Dyes can be used in solid, liquid, or gas phases. Liquid solutions of dyes are especially convenient, because the concentration, and hence their absorption and gain, is readily controlled in them.

The spectral range of flashlamp-pumped dye lasers extends presently from 340 to 850 nm. Their spectra show greater variety than those of laser-pumped dye lasers.

A rough selection of the dye-laser emission wavelength is possible by choice of the dye, the solvent, and the resonator Q, see Schafer.[55] Fine tuning and simultaneous attaintment of small line widths can be achieved by using wavelength-selective resonators with wavelength-selective devices as follows:

1. Gratings and prisms for angular wavelength separation.
2. Interferometric wavelength discrimination, using more often Fabry–Perot etalons.
3. Rotational wavelength separation which uses birefringent filters. Brewster plate polarizers, etc.
4. Resonators with wavelength-selective distributed feedback.

TABLE II. Some Commercial Continuous-Wave Gas Laser Characteristics

Laser	Color	Wavelength emission nm	Power (mW)	Divergence (m-rad)	Beam diameter (mm)
He–Ne	Green	543	0.2	0.92	0.75
	Red	633	0.1–50	0.5–6	0.5–15
	IR	1150	1–15	0.8–2	0.8–2
	IR	3390	0.7–3	3–6	0.8–2
He–Cd	UV	325	1–10	0.5–1.5	0.7–2.0
	Blue	442	8–50	0.6–1.5	0.6–1.5
Ar-Ion	UV-Green	330–530	$15-4\times10^4$	0.5–1.5	0.7–1.6
Kr	UV-IR	337–860	$250-7\times10^3$	0.5–1.3	1–2
CO_2	IR	1.06×10^4	$10-10^5$	1–10	1.4–10
	IR	1.06×10^4	10^5-10^6	1–5	5–25
	IR	1.06×10^4	$10^6-2\times10^7$	1–5	10–75
H-cyanide	IR	3.11×10^4			
		33.7×10^4	$10-10^3$	40	10
Methyl fluoride	IR	$9.6\times10^4-18\times10^5$	$10-1^{-2}$	40	1–3
Water	IR	11.8×10^4			
	IR	7.8×10^4			
	IR	2.8×10^4	1–10	10	5
Hydrogen fluoride	IR	2.58×10^3			
	IR	3.05×10^3	$2\times10^3-35\times10^3$	1	1.5

TABLE III. Commercial Pulsed Gas Laser Characteristics

Laser	Wavelength (nm)	Pulsed energy (J)	Divergence (m-rad)	Beam diameter (mm)
Argon ion	458	5×10^{-8}	0.7–5	1.2–2.5
	528	3×10^{-5}	0.7–5	1.2–2.5
Carbon dioxide	10.6×10^3	$3\times10^{-5}-75$	0.6–10	2–90
		10^2-10^3	0.5–2	8–200
		10^3-10^4	0.3–0.5	200–350
Carbon monoxide	$5\times10^3-6\times10^3$	$8\times10^{-3}-2.2$	0.6	5–32
Copper vapor	511	$10^{-4}-2\times10^{-4}$	1	15–35
Deuterium fluoride	$3.5\times10^3-4\times10^3$	0.04–0.7	0.5–0.75	10–30
Hydrogen cyanide	$31.1\times10^4-33.7\times10^4$	10^{-3}	40	10
Hydrogen fluoride	$(2.8-3.2)10^3$	$5\times10^{-3}-1$	0.3–4	25–30
Krypton	458–647	$5\times10^{-8}-10^{-6}$	0.7	1.2
Nitrogen	337	$5\times10^{-6}-10^{-1}$	0.1–10	$1-6\times32$
Nitrous oxide	$(5-6)10^3$	0.3–2.2	0.6	30
Water	$(11.8, 1.8, 2.8)10^4$	10^{-5}	15	8
Xenon fluoride	355	$(5-10)10^{-2}$	4	$4\times15-20$
Argon fluoride	193	$(4-10)10^{-2}$	4	$4\times15-20$
Krypton fluoride	249	$3\times10^{-5}-0.25$	3–4	3–20

TABLE IV. Standard Specifications for Dye Lasers

Power output	2 W
Tuning range	420–710 nm
Efficiency (Optical power out)	0.35%
Excitation power in mode locked pulse-width	$0.5 \times 10E-12$ sec

As in the lasers just described, the dye can lase in both, continuous-wave (CW) and pulsed ways. One standard set of specifications for a CW dye laser is shown in Table IV.

This laser is usually driven by using other lasers such as Nd-glass and ruby lasers obtaining very high power pulses. Some of the more often used dyes and their wavelengths are given in Table V.

5.5.6. Solid-State Lasers

In recent years, the field of the solid-state laser has made rapid advances in the quality and variety of materials available, as well as in the design and operation of sophisticated laser systems for many applications. An example is the Nd-glass laser operating at a wavelength of 1.06 μm, which is used to investigate the physics of inertial confinement, laser fusion, and in many other industrial scientific, and military applications.

The intent of this section is to provide an overview of solid-state lasers as light sources. Many other properties can be found in the chapter by Brown.[56]

The main classes of solid-state lasers which are most widely used are: Rubi laser, Neodymium glass, Neodymium YAG, and semiconductor lasers.

1. *Rubi Lasers.* The rubi laser was the first operational laser medium. The laser transition observed corresponds to 649.3 nm in the red. To excite this laser, a flash lamp is generally used. Since the absorption band extends from the visible into the ultraviolet, most helical flash lamps used for this laser are quartz-enveloped. They are operated with more than 100 amp/cm^2 current densities, supplied by a capacitor bank which is energized to a voltage lower than the breakdown voltage of the flash lamp. Then, a high-voltage trigger pulse of about 20 kV fires the lamp. Because the very high current densities produce a high temperature in the rod, air or liquid coolants are used in commercial systems.

One of the cavity configurations which offers several advantages is a rod cut at the Brewster angle and mounted with external mirrors. This configuration allows external devices to be placed inside the laser resonator and also allows polarized output. Since the oscillation produces multimodal

TABLE V. Some of the Dye Lasing Wavelengths (Flashlamp-pumped)

Dye	Solvent	Lasing Wavelength (nm)
Rhodamine 110	HFIP	540
	TFE	550
	EtOH, basic	560
	EtOH, acidic	570
	DMSO	575
Rhomadine 19	EtOH, basic	575
	EtOH, acidic	585
Rhodamine 6G	HFIP	570
	TFE	575
	EtOH	590
	DPA	595
	DMSO	600
Rhodamine B	EtOH, basic	610
	EtOH, acidic	620
Rhodamine 3B	HFIP	610
	TFE	610
	E OH	620
	DMSO	630
Rhodamine 101	HFIP	625
	TFE	625
	EtOH, basic	630
	EtOH, acidic	640
	DMSO	650

HFIP Hexafluoroisopropanol
TFE Trifluoroethanol
DMSO dimethyl sulfoxide
DPA N,N-Dipropylacetamide
An extensive list of laser dyes can be found in Refs. 14 and 15.

frequencies of transverse modes, the oscillator can be forced to operate in the TEM_{00} mode with the insertion of an aperture of about 2 mm into the resonator. This results in a Gaussian beam intensity profile and a uniphase wave front. It is well known that the rubi laser is a pulsed laser. Some techniques to obtain multiple pulses from such a laser depend on the time separation between the pulses. The more common pulse separation rates used in holography are 1 μs to 1 ms for double-pulsed holography. There is another technique which allows one to reduce the pulse separation, giving a pulse separation of 20 ns to 1 ms, as described by Koechner.[57] The coherence length of these rubi lasers depends on the pulse separation and ranges from 0.5 to 10 m.

2. *Nd-YAG-Laser.* This laser consists of a rare-earth ion Nd_3^+ incorporated into a YAG host. The laser transition occurs at 1064 nm. The linewidth is homogeneously broadened by thermal lattice vibrations. Inserting a prism or other dispersive element into the laser cavity, it is possible to force the laser to oscillate on one of the more than twenty transitions observed in Nd-YAG.

3. *Nd-glass.* As mentioned elsewhere, Nd-glass has been the material for most systems built for high-power applications; the laser emission is in the 1064 nm range. In some applications like pumping dye lasers and satellite ranging, it is common to use a KDP (potassium dihydrogen phosphate) frequency doubler to obtain visible light at 532 nm, with low beam divergence (Nathan[58]).

5.5.7. Semiconductor Lasers

In comparison with gas or other solid-state lasers, the diode lasers suffer mainly from a much broader far-field pattern, the present nonexistence of high-efficiency, truly visible lasers at room temperature (either CW or pulsed), and the lack of coherence. However, in some cases the coherence of a laser is so good that it can substitute for a gas laser (He–He) with some advantage, as shown by Yoshikawa.[59] The laser diode is a compact, efficient, easily modulated, and low-cost radiation source. For applications where coherence requirements are not that important, the beam can be focused with a very modest lens system. It is also well suited as a light source for fiberoptic communication systems. Furthermore, the devices are sufficiently efficient to permit their use in communication systems where bandwidths in the MHz range are needed. Usually, room-temperature semiconductor lasers are pulsed. In order to take advantage of the fast rise time that is possible with the semiconductor laser, specially designed pulse circuits must be used. These circuits are similar to the ones used with the solid-state flash lamp. The peak current obtained with such a circuit may be of the order of 50 amperes, with a pulse width of 100 nanoseconds, producing a peak laser power of approximately of 10 watts.

The onset of lasing of a light-emitting diode is not as dramatic as that of a typical gas laser. As the current of an LED is increased to threshold, the output radiance increases abruptly in the direction of the laser beam. The beam divergence angle decreases, and the spectral width of the emitted light narrows greatly. The threshold current for laser operation is highly dependent on the temperature of the diode, between room temperature ($\sim 300°K$) and the temperature of liquid nitrogen ($77°K$). The threshold current decreases by more than one order of magnitude. Cooling increase the radiative efficiency of the GaAs diode laser by more than seven times its room-temperature value and permits a large average output power, since the percentage of the time the laser is on can be increased by a factor of 40. In addition, cooling increases the band gap of GaAs shifting the output from the 900 nm to the 850 nm region where detectors are more sensitive (Kressel and Ettenberg[60]).

All the emission occurs in the narrow junction region. Because of the confinement of the beam to a very small region, diffraction of the light results in a large beam divergence. Light of wavelength λ passing through a slit of width a will be diffracted in a cone of light with an angular radius θ given by

$$\sin \theta = \frac{\lambda}{a}. \tag{5.3}$$

For a typical semiconductor laser, such as GaAs, $\lambda = 900$ nm, and $a = 6$ μm, $\theta = \arcsin 0.15 = 8.6°$. This angular spread is quite large when compared to that of other types of lasers. The fan beam is characteristic of all semiconductor lasers. Semiconductor lasers that are fabricated from a single semiconductor (e.g., GaAs) are called *homojunction lasers*; and if they are fabricated with more than one material, they are called *heterojunction lasers* and have some advantages compared with the former. Much higher efficiency and a much lower threshold-current density is obtained in heterojunction. These heterojunction lasers have the advantage that the current density threshold is reduced, and there are two types, single and double heterostructure, as shown in Table VI.

Some commercially available heterostructure-junction lasers operate continuously at room temperature. One such laser has an output at 820-nm wavelength with 5 mW of power, and a threshold current of less than 300 mA, as reported by O'Shea *et al.*[53] They are commercially available with powers up to 50 mW for a single emitter with multilongitudinal modes and can have up to 2.6 W for a phased array, see Stern.[61] In a single-emitter, single-longitudinal mode configuration, they can have a CW output of 15-20 mW. Their lifetimes are more than 10,000 hours.

Diode lasers may operate either in a pulsed mode, by modulating the current flowing through the diode, or continuously (CW) with a dc bias.

TABLE VI. Current Threshold for Semiconductor Lasers

Homostructure	40,000 amperes/cm^2
Single-Heterostructure	8,000 amperes/cm^2
Double-Heterostructure	2,000 amperes/cm^2

TABLE VII. Some Characteristics of Semiconductor Lasers

Laser	Wavelength in nm
ZnO	370
CdS	490
AlGaAs	620–900
GaAsP	610–900
GaAs	830–920
ZnS	330
ZnCdS	330–490
ZnSe	460
ZnTe	530
GaSe	590–600
CdSSe	590–690
CdSe	690
CdTe	785
InGaP	590–900
InP	890–910
GaAsSb	950–1600
$CdSnP_2$	1100
InAsP	900–3200
GaSb	1500–1600
InGaAs	850–3200
Cd_3P_2	2100
InAs	3000–3200
InAsSb	1000–5300
CdHgTe	3800–4100
Te	3720
PbS	4300
InSb	4800–5300
PbTe	6500
PbSSe	4700–5500
PbSe	8500
PbSnTe	6500–32000
PbSnSe	8500–32000

Notes:
(1) The first four lasers work at room temperature.
(2) The far-field emission pattern in the plane of the junction typically has an angular spread of 15° parallel to the junction and 3° perpendicular to the junction. This asymmetry in the two divergences leads to an elliptical light spot.

For CW operation, the diode output usually has multiple longitudinal modes. However, a single-longitudinal mode can be obtained by a proper design of the waveguide geometry. Some other characteristics are given in Table VII.

The coherence length in a laser diode is measured with a modified Twyman–Green interferometer, using the laser as a source. The fringe visibility is very good for path-length differences up to 15 m, as shown by Creath.[62]

References

1. W. Elenbaas, *Light Sources*, Crane, Russak and Company, New York, 1972.
2. Illuminating Engineering Society, *I.E.S. Lighting Handbook*, 5th ed., Illum. Eng. Soc., New York, 1972.
3. J. M. Anderson and J. S. Saby, "The Electric Lamp: 100 Years of Applied Physics," *Physics Today* **32**, 33 (1979).
4. F. E. Carlson and C. N. Clark, "Light Sources for Optical Devices," in *Applied Optics and Optical Engineering*, Vol. 1, Chap. 2, Academic Press, New York, 1975.
5. G. J. Zissis and A. J. Larocca, "Optical Radiation and Sources," in *Handbook of Optics*, Sec. 3, (W. G. Driscoll, Ed.), McGraw-Hill, New York, 1978.
6. F. Kreith and J. F. Keider, *Principles of Solar Engineering*, McGraw-Hill, New York, 1978
7. Iqbal Muhammad, *An Introduction to Solar Radiation*, Academic Press, New York, 1983.
8. N. J. Kreidl and J. L. Rood, "*Optical Materials*," in *Applied Optics and Optical Engineering*, Vol. 1, Chap. 5 (R. Kingslake, ed.), Academic Press, New York, 1965.
9. W. L. Wolfe, "Properties of Optical Materials," in *Handbook of Optics*, Chap. 7 (W. G. Driscoll, ed.), McGraw-Hill, New York, 1978.
10. R. D. Larrabee, "Spectral Emissivity of Tungsten," *J. Opt. Soc. Am.* **49**, 619 (1959).
11. W. E. Forsythe and E. Q. Adams, "Radiating Characteristics of Tungsten and Tungsten Lamps," *J. Opt. Soc. Am.* **35**, 108 (1945).
12. J. E. Eby and R. E. Levin, "Incoherent Light Sources," in *Applied Optics and Optical Engineering*, Vol. 7, Chap. 1, Academic Press, New York, 1979.
13. R. W. Boyd, *Radiometry and the Detection of Optical Radiation*," Chap. 2, Wiley, New York, 1983.
14. W. E. Tolles and L. C. Ihring, "Modulation of Radiant Energy from Incandescent Filaments," *J. Opt. Soc. Am.* **47**, 101 (1957).
15. R. D. Kingslake, *Optical System Design*, Academic Press, New York, 1983.
16. F. J. Studer and R. F. VanBeers, "Modification of Spectrum of Tungsten Filament Quartz-Iodine Lamps due to Iodine Vapor," *J. Opt. Soc. Am.* **54**, 945 (1964).
17. R. Stair, W. E. Schneider, and J. K. Jackson, "A New Standard of Spectral Irradiance," *Appl. Opt.* **2**, 1151 (1963).
18. F. Grum and R. J. Becherer, "Optical Radiation Measurements," in *Radiometry*, Vol. 1, Chap. 5, Academic Press, New York, 1979.
19. General Electric Co., *Catalog P3-61P: Photographic Lamp & Equipment Guide*, Cleveland Ohio.
20. V. J. Francis, *Fundamentals of Discharge Tube Circuits*, Methuen, London, 1948.
21. R. E. Pulfrey, "Optically Stabilized Power Supply for Mercury Arc Lamp far IR Sources," *Appl. Opt.* **15**, 308 (1976).
22. J. P. Latil, "A Fully Automatic Continually Operating, Very High-Intensity, Carbon Arc Lamp," *J. Opt. Soc. Am.* **44**, 1 (1954).

23. M. R. Null and W. W. Lozier, "Carbon Arc as a Radiation Standard," *J. Opt. Soc. Am.* **52**, 1156 (1962).
24. Canrad-Hanovia Inc., *Compact Arc Lamps, Cat. No.* 0642/181, Newark, New Jersey.
25. W. Budde, "Stability of High-Pressure Xenon Lamps," *J. Opt. Soc. Am.* **52**, 343 (1962).
26. D. C. Pritchard, *Lighting*, 2nd ed. Longmans, London, 1978.
27. J. P. Frier and M. E. Frier, *Industrial Lighting Systems*, McGraw-Hill, New York, 1980.
28. W. T. Welford, *Optics*, 2nd ed., Oxford Physics Series, Oxford University Press, Oxford, 1981.
29. A. A. Michelson and E. W. Morley, "On the Feasibility of Establishing a Light Wave as the Ultimate Standard of Length," *Am. J. Sci.* **38**, 181 (1889).
30. W. F. Meggers, "A Light Wave of Artificial Mercury as the Ultimate Standard of Length," *J. Opt. Soc. Am.* **38**, 7 (1948).
31. Oriel Corp., *Catalog Vol. 2, Light Sources Monocromators and Detection Systems*, Stratford, Conn., 1985.
32. Ealing Corp., *Ealing Optics Complete Catalog*, 84/85, p. 242, South Natick, Mass., 1985.
33. B. A. Lengyel, *Lasers*, 2nd ed. Chap. 4, Wiley-Interscience, New York, 1971.
34. I. S. Marshak, "Limiting Parameters and Generalized Characteristics of Xenon Lamps," *Appl. Opt.* **2**, 793 (1963).
35. R. H. Dishington, W. R. Hook, and R. P. Hilberg, "Flashlamp Discharge and Laser Efficiency," *Appl. Opt.* **13**, 2300 (1974).
36. T. Efthymiopoulos, and B. K. Garside, "High-Energy Short-Pulse Flashlamps: Operating Characteristics," *Appl. Opt.* **16**, 70 (1977).
37. J. F. Holzrichter and J. L. Emmett, "Design and Analysis of a High Brightness Axial Flash Lamp," *Appl. Opt*, **8**, 1459 (1969).
38. M. A. Gushinov, "Spectral Enhancement of Near UV Xenon Flashlamps," *Appl. Opt.* **14**, 2645 (1975).
39. R. H. Dishington, "Flash Lamp Drive Circuit Optimization for Lasers," *Appl. Opt.*, **16**, 1578 (1977).
40. Y. Levy, G. Neumann, and D. Treves, "Ablative Flashlamps for High Peak Power Dye Lasers," *Appl. Opt.* **16**, 2293 (1977).
41. General Instrument Corp., *Catalog of Optoelectronic Products* 1983, Optoelectronic Division, 3400 Hillview Ave., Palo Alto, Calif. 94304.
42. Hewlett-Packard Optoelectronics Division, *Optoelectronics/Fiber-Optics Applications Manual*, 2nd ed., McGraw-Hill, New York, 1981.
43. Motorola Semiconductor Products Inc., *Optoelectronic Device Data*, Second Printing, 1981, Box 20912, Phoenix, Arizona 85036.
44. S. M. Sze, *Physics of Semiconductor Devices*, Chap. 12, Wiley, New York, 1969.
45. T. Kawabata, T. Matsuda, and S. Koike, "GaN Blue Light Emitting Diodes Prepared by Metalorganic Chemical Vapor Deposition," *J. Appl. Phys*, **56**, 2367 (1984).
46. J. Jarominsky, "Optical Efficiency of LEDs with Hemispherical Lenses," *Appl. Opt.* **21**, 2461 (1982).
47. W. S. Holden, W. M. Hubbard, and S. D. Personick, "Chromatic Delay in Light Emitting Diodes," *Appl. Opt*, **13**, 1050 (1974).
48. J. Pucilowsky, R. Schuman, and J. Velasquez, "Contrast Enhancement of Light Emitting Diode Displays," *Appl. Opt*, **13**, 2248 (1974).
49. R. W. Dawson, "Pulse Widening in a Multimode Optical Fiber Exited by a Pulsed GaAs LED," *Appl. Opt*, **13**, 264 (1974).
50. M. R. Allyn, R. W. Dixon, and R. Z. Bachrach, "Visibility of Red and Green Electroluminescent Diodes for Color-Anomalous Observers," *Appl. Opt.* **11**, 2450 (1972).
51. J. Svetlik, "Simple Methods for the Measurement of Laser Beam Parameters," *Appl. Opt*, **13**, 1276 (1974).

52. Y. Li and E. Wolf, "Focal Shifts in Diffracted Converging Spherical Waves," *Opt. Comm,* **39**, 211 (1981).
53. D. C. O'Shea, W. R. Collen, and W. T. Rhodes, *Introduction to Lasers and their Applications,* Chap. 6, Addison-Wesley, Reading, Massachussetts, 1978.
54. P. W. Smith, "Single-Frequency Lasers," in *Lasers* (A. K. Levine and A. J. DeMaria, eds.), vol. 4 Marcel Dekker, 1976.
55. F. P. Schafer, "Dye Lasers," in *Topics in Applied Physics,* 2nd ed., Vol. 1, Springer-Verlag, Berlin, 1977.
56. D. C. Brown, "Solid State Lasers," in *Applied Optics and Optical Engineering,* Vol. 6, Chap. 1, Academic Press, New York, 1980.
57. W. Koechner, "Equipment and Procedures," in Handbook of Optical Holography, Chap. 8 (H. J. Caulfield, ed.), Academic Press, New York, 1979.
58. M. I. Nathan, "Semiconductor Lasers," in *Laser Devices and Applications* (I. P. Kaminow and A. E. Siegman, eds.), IEEE Press, 1973.
59. M. Yoshikawa, N. Suzuki, and T. Suzuki, "Coherence Characteristics of a Single-Mode GaAlAs Laser Diode," *J. Opt. Soc. Am.* **71**, 171 (1981).
60. H. Kressel and M. Ettenberg, "Semiconductor Lasers," in *Handbook of Laser Science and Technology;* Vol. 1, *Lasers and Masers* (M. J. Weber, ed.), CRC Press, Boca Raton, Florida, 1981.
61. F. Stern, "Semiconductor Lasers; Theory," in *Laser Handbook* Vol. 1 (F. T. Arecchi and E. O. Schulz-Dubois, eds.), North-Holland Publ., Amsterdam, 1972.
62. K. Creath, "Interferometric Investigation of a Diode Laser Source," *Appl. Opt.* **24**, 1291 (1985).

6. OPTICAL FILTERS

Fred Goldstein

FTG Software
P.O. Box 358
Chatham, New Jersey 07928

6.1. Introduction

Optical filters are devices which alter the spectral properties of incident light. A comprehensive overview of a wide variety of coatings and filters has been given by Dobrowolski.[1] The discussion in this chapter is limited to the subset of devices generally regarded as within the technological range of commercial suppliers. Most examples involve visible transmission filters at normal incidence.

What has changed since Dobrowolski's 1978 review article (actually written in 1970)? While fundamental theories and properties of materials are essentially the same, continuing developments in vacuum technology and instrumentation make it possible to fabricate a wider variety of multilayer devices. Filters which were once state-of-the art are now commonplace.

The task of designing filters has become simpler through recent developments in computer hardware and software. Thin-film calculations formerly requiring mainframes and resident programming experts are now performed with powerful desktop computers and commercially available software packages. Technicians armed with such capability can often arrive (perhaps by brute force) at thin-film constructions as useful as those formerly produced only with expert understanding.

This chapter emphasizes the *types* of computations that can aid in selecting filters. Users should expect manufacturers to perform similar analyses in meeting their exact requirements. An annotated software program FILTER is included for those interested in the thin-film design process. This is followed by a discussion of selection criteria and several sections illustrating the general properties of various filter types. Although examples are restricted to the visible region, general principles apply to the ultraviolet (UV) and infrared (IR) as well. Finally, a section on thin-film deposition is included as an aid in evaluating the capabilities of coating suppliers.

6.2. Multilayer Thin Films: Matrix Theory

Optical filters may be constructed from absorbing glasses and/or multilayer thin-film interference devices composed of dielectric and/or metallic layers. The properties of colored glasses are given in manufacturers' catalogs. Unless there is a large requirement where a special melt is justified, the user is generally limited to combinations of standard filters in varying thicknesses.

The situation with regard to multilayer devices is far more complex. A distinction must be made between the virtually unlimited number of filters which are theoretically possible and the rather smaller number which may be actually manufactured. While absorbing glasses are used in transmission, multilayers can also be used in reflection.

The relationships describing the reflection and transmission of optical thin-film filters are derived from Maxwell's equations. Following Macleod's text,[2] we represent the characteristic matrix of a thin-film assembly, illustrated in Fig. 1, of nonabsorbing isotropic layers at normal incidence as the product of separate matrices for each layer:

$$\begin{pmatrix} B \\ C \end{pmatrix} = \left\{ \prod_{j=1}^{N} \begin{pmatrix} \cos(\delta_j) & i\sin(\delta_j)/n_j \\ in_j \sin(\delta_j) & \cos(\delta_j) \end{pmatrix} \right\} \begin{pmatrix} 1 \\ n_{\text{sub}} \end{pmatrix}, \qquad (6.1)$$

where

$$\delta_j = \frac{2\pi n_j}{\lambda} d_j. \qquad (6.2)$$

Complex reflection and transmission amplitudes of the assembly are given by

$$r = \frac{(n_0 B - C)}{(n_0 B + C)} \qquad t = \frac{2n_0}{(n_0 B + C)}. \qquad (6.3)$$

Reflectance and transmittance are the measurable quantities of interest given by

$$R = |r|^2 \qquad T = \text{Re}(n_{\text{SUB}})|t|^2/n_0. \qquad (6.4)$$

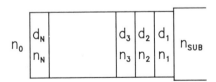

FIG. 1. Multilayer filter. Note that layer 1 is next to the substrate.

6.2.1. Including the Substrate

The above discussion is based on multiple internal addition of complex amplitudes: layers are "thin" in comparison with the coherence length of light. This is quite valid within multilayer devices where film thicknesses are of the order of a quarter wavelength of light. In this chapter, substrates are usually regarded as "thick," and intensity summations (neglecting phase) are appropriate. The validity of this assumption is discussed in a later section on band-pass filters.

The above can be extended to nonnormal incidence by replacing n_j with $n_j \cos \theta_j$ for S polarization and $n_j/\cos \theta_j$ for P polarization. For absorbing films, n_j is replaced with $n_j - ik_j$, where k is a measure of absorption in the medium; trigonometric functions are replaced with their complex equivalents, etc.

The optical properties of an absorbing filter glass substrate are described by n and k. The internal transmittance $t(\lambda)$ represents the diminution in intensity of light waves propagating through the substrate. Starting with the plane wave solution to Maxwell's equations, it can be shown that

$$t = \exp\left(\frac{-4\pi k d}{\lambda}\right). \tag{6.5}$$

If t_1 and t_2 are the transmittances for the thicknesses d_1 and d_2, respectively, we find that:

$$t_1 = t_2^{(d_1/d_2)}. \tag{6.6}$$

This expression enables one to calculate the internal spectral response of a filter glass as the thickness of glass is varied. This becomes more useful when we relate internal transmittance to total transmittance as measured in a spectrophotometer.

Referring to Fig. 2, consider the general case of a film deposited on a substrate. Rf is the reflectance at the filmed surface, while Rs is the corresponding value for the uncoated surface. Rs is given by the Fresnel

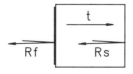

FIG. 2. Film on absorbing substrate with internal transmittance t. Rf is the reflectance from the filmed surface, Rs from the uncoated substrate surface.

expression

$$\text{Rs} = \left(\frac{1-n_{\text{SUB}}}{1+n_{\text{SUB}}}\right)^2. \tag{6.7}$$

Based on incoherent addition of intensities, total transmittance T is related to internal transmittance t by

$$T = \frac{t(1-\text{Rf})(1-\text{Rs})}{1-t^2(\text{Rf}\cdot\text{Rs})}. \tag{6.8}$$

This expression is used in the BASIC multilayer program FILTER given below. While manufacturers' catalog values for internal transmittances can be used for initial calculations, more accurate determinations may be required. If the substrate is uncoated, we set Rf = Rs and arrive at expressions which relate internal and total transmittance. Note that these expressions are based on knowledge of refractive index n.

Optical density is a quantity frequently utilized in the specification of blocking filters. This is related to transmittance by

$$\text{OD} = -\log_{10}(T). \tag{6.9}$$

This definition is not strictly correct in that it really applies to internal transmittance. In this chapter, we adopt a loose definition and apply this quantity to all types of filters.

6.2.2. BASIC Program FILTER

The program given below is an implementation of the above equations for nondispersive (index n constant as a function of wavelength) nonabsorbing films deposited on a slightly absorbing substrate at normal incidence. As substrate $k \ll n$, it may be neglected in the multilayer interference calculation.

Notation in optical thin films is not standard and can lead to confusion. Most theoretical articles assume that layer 1 adjoins the entrance medium (air). On the other hand, coating technicians sensibly assume that layer 1 is the first layer on the substrate—an assumption in accordance with this chapter. As illustrated in Table I, the thickness of a layer can be described by various quantities.

In the listed program, layers are described by Q and n. Many published articles use K for layer thickness: constructions such as [HLH] are expressed here as [.25H .25L .25H] where "L" and "H" indicate low and high index films. The design wavelength "Wdes" specifies the wavelength at which

MULTILAYER THIN FILMS: MATRIX THEORY 277

TABLE I. Thickness Quantities for Thin-Film Layers

physical thickness	d
optical thickness	nd
relative optical thickness	$Q = nd/\text{Wdes}$
phase thickness	$2\pi Q$
number of quarter waves	$K = 4Q$

```
1000   !   ------------------------------------------------
1010   !
1020   !                   Program FILTER
1030   !
1040   !   Multilayer Dielectric on Absorbing (k ≪ n) Filter Glass
1050   !
1060   !
1070   !   ------------------------------------------------
1080   INTEGER Ktype
1090   RAD                 ! Radians
1100   !
1110   !   ------------------------------------------------
1120   !   Nsub .......... Substrate index n
1130   !   Thick0 ......... Reference substrate thickness in mm. Thickness
1140   !                    for known t (catalog) or %T (measured).
1150   !   Thick .......... Actual substrate thickness in mm.
1160   !   Ktype .......... 0 Nonabsorbing substrate
1170   !                    1 Reference substrate internal transmittance t
1180   !                      (catalog value) known
1190   !                    2 Reference (uncoated) substrate measured %T known
1200   !   Wdes .......... Design wavelength in nm.
1210   !
1220   READ Nsub,Thick0,Thick,Ktype,Wdes
1230   DATA 1.52,  1.0,  2.0,  2,  550.
1240   !
1250   !   ------------------------------------------------
1260   !
1270   Rs=((1.-Nsub)/(1.+Nsub))^2    ! Substrate side 2 reflectance
1280   Ts=1.-Rs                       ! Rs=0, Ts=1. if AR coat on side 2
1290   Tmax=100.*Ts*Ts/(1.-Rs*Rs)    ! Max %T for a nonabsorbing substrate
1300   !                              with index Nsub
1310   !
1320   !   ------------------------------------------------
1330   !
1340   !   Design format (starting at substrate):
1350   !              Q1,N1,  Q2,N2,  Q3,N3,  Q4,N4,..., 0
1360   !
1370   !   QJ ........... Optical thickness/Wdes for layer J (1 QWOT=.25)
1380   !                  0 (zero) ends design
1390   !   NJ ........... Index of refraction for layer J
1400   !
```

```
1410    Design:    !  Follow with as many DATA statements as required
1420               !  DATA 0 can be used for uncoated filters
1430               !
1440    DATA  .25,  2.3,  .25,  1.46,  .25,  2.3,  .25,  1.46
1450    DATA  .25,  2.3,  .25,  1.46,  .25,  2.3,  0
1460    !
1470    !   ------------------------------------------------
1480    !
1490    !
1500    LOOP
1510       INPUT "Evaluation wavelength in nm? (0 to QUIT)", Weval
1520       EXIT IF Weval<=0
1530       R1=1.
1540       R2=Nsub
1550       R3=0
1560       R4=0
1570       RESTORE design
1580       LOOP
1590          READ Q            ! Optical thickness (1 QWOT=.25)
1600          EXIT IF Q=0
1610          READ N            ! Index
1620          G=2.*PI*Wdes*Q/Weval
1630          P1=COS(G)
1640          Sg=SIN(G)
1650          P2=Sg/N
1660          P3=Sg*N
1670          A1=P1*R1−P2*R4    ! Update characteristic matrix
1680          A2=P1*R2−P3*R3
1690          R3=P1*R3+P2*R2
1700          R4=P3*R1+P1*R4
1710          R1=A1
1720          R2=A2
1730       END LOOP
1740       X1=R1−R2
1750       X2=R3−R4
1760       Y1=R1+R2
1770       Y2=R3+R4
1780       D=Y1*Y1+Y2*Y2
1790       Rreal=(X1*Y1+X2*Y2)/D
1800       Rimag=(X2*Y1−X1*Y2)/D
1810       Refl=Rreal*Rreal+Rimag*Rimag    ! Reflectance
1820       Trns=1.−Refl
1830       SELECT K type
1840       CASE 0
1850          Tint=1.           ! Substrate internal transmittance=1
1860       CASE 1
1870          INPUT "REFERENCE substrate internal transmittance ? (0−1)",Tint
1880          IF Tint<0 OR Tint>1 THEN
1890             PRINT "ILLEGAL ENTRY (<0 or >1)"
1900             BEEP 200,.2
1910             GOTO 1870
```

1920		END IF
1930		IF Tint>0 THEN Tint=Tint^(Thick/Thick0)
1940		CASE 2
1950		INPUT "REFERENCE substrate % transmittance ?",Tper
1960		If Tper>Tmax OR Tper<0 THEN
1970		PRINT "ILLEGAL ENTRY (<0 or > "&VAL$(DROUND(Tmax,8))&"%)"
1980		BEEP 200.,2 ! Cannot exceed max theoretical %T
1990		GOTO 1950 ! for an uncoated substrate
2000		END IF
2010		Tint=.01*Tper
2020		IF Tint>0 THEN
2030		B=Ts*Ts
2040		IF Tint>.001 THEN
2050		Tint=(SQR(B*B+4.*(Rs*Tint)^2)−B)/(2.*Rs*Rs*Tint)
2060		ELSE
2070		Tint=Tint/B ! Avoid math error for Tint near 0
2080		END IF
2090		Tint=Tint^(Thick/Thick0)
2100		END IF
2110		END SELECT
2120		Trns=100.*Tint*Trns*Ts/(1.−Tint*Tint*Refl*Rs)
2130		PRINT "Wavelength ";Weval;"nm T= "&VAL$(DROUND(Trns,5))&"%"
2140		END LOOP
2150		PRINT
2160		PRINT "**** PROGRAM ENDED − RUN to RESTART ****"
2170		!
2180		END
3000	!	--
3010	!	
3020	!	EXAMPLES
3030	!	
3040	!	Use the above sample DATA and enter %T (Ref)=50% for each
3050	!	wavelength to obtain the following:
3060	!	
3070	!	Wavelength 400 nm T=21.249%
3075	!	Wavelength 500 nm T=3.2042%
3080	!	Wavelength 600 nm T=2.7971%
3090	!	Wavelength 700 nm T=16.017%
3100	!	Wavelength 800 nm T=24.968%
3110	!	
3120	!	
3130	!	Use the above sample DATA except change Ktype to 1. Enter
3140	!	t (Ref)=0.8 for each wavelength to obtain the following:
3150	!	
3160	!	Wavelength 400 nm T=45.917%
3165	!	Wavelength 500 nm T=6.9850%
3170	!	Wavelength 600 nm T=6.0988%
3180	!	Wavelength 700 nm T=34.699%
3190	!	Wavelength 800 nm T=53.857%
3200	!	
4000	!	--

$Q = .25$ is a quarter-wave optical thickness. This is useful because the entire spectrum can be shifted to lower or higher wavelengths by simply decreasing or increasing Wdes.

Various parameters including thin-film construction, substrate index "Nsub" and thickness "Thick" are input as DATA statements. "Thick0" is the thickness of a reference sample for which internal transmittance t (0–1) or measured transmittance % T (0–100) is known. The program prompts the user to input t or % T according to variable "Ktype". Uncoated substrates may be evaluated by setting the thickness of the first layer to zero.

The program is written in Hewlett-Packard HP 9000 BASIC. It is easily modified for Pascal, FORTRAN, or other BASICs. Note that LOOP... END LOOP constructs can be replaced with GOTO statements, and SELECT... END SELECT constructs can be replaced with IF... THEN statements.

6.3. Selecting Filters

The large variety of possible filters includes short-pass, long-pass, band-pass, rejection, spectral-conversion, and neutral-density filters. These are discussed in further detail below. While satisfactory filters of one type may be available off the shelf, the production of other types may stretch the state of the art.

Filter specifications are ultimately based on system performance and involve figures of merit such as signal-to-noise, CIE (International Commission on Illumination) coordinates, photopic response, detector output, etc. In many instances, filter requirements can be expressed as transmittance specifications: (1) Minimum, maximum, average values in given wavelength regions, etc. (2) Cut wavelength (typically at T = 50% or half maximum), peak wavelength, half-width, etc.

When system performance is a complex function of filter response, it is the *responsibility of the system designer* to arrive at a set of criteria which are mathematically related to spectral performance. Difficulties can arise when users specify "color temperature" and suppliers think "transmittance versus wavelength." Translation of complex criteria into simple transmittance specifications can overspecify a filter and result in lower manufacturing yields and consequent higher costs.

Overspecification can also result from insufficient system analysis. The author has seen coatings specified as "less than 0.5% reflectance from 15° to 45° angle of incidence," when the system designer was really interested in maximum light throughput. While the amount of light from 40° to 45°

might be 10% of the amount between 25° to 30°, the <0.5% specification applies equally to both ranges! Such parts might fail incoming inspection but perform adequately in systems.

Depending on the sophistication of his product line, the filter manufacturer should have access to computer facilities for evaluating combinations of filters with or without dielectric or metal coatings. Such programs might range from the one included in this chapter to a comprehensive optimizing package such as FILM∗STAR.[3] This package allows virtually any type of complex specification through the ability to pass calculated values to supplementary subprograms. It also provides the means to calculate the response of a series of reflection and transmission filters.

Assuming equivalent system performance, it seems reasonable to select absorbing glass filters over more expensive multilayer filters. (A glass filter with only antireflection coatings is classified as an absorbing filter.) Once a melt is characterized by grinding and polishing reference samples, spectral performance should depend solely on thickness. Of course, continuous testing is required—especially if there are indications of variations within a melt.

The major manufacturers of colored glasses include Corning, Hoya, and Schott. The Schott catalog *Optical Glass Filters* is particularly informative. A number of optical shops specialize in cutting, polishing, coating, and cementing colored glasses to order. These shops can be located under "Filters" in buyers' guides such as those published by Laser Focus or Photonics Spectra. Antireflection (AR) coatings can range from a single layer of magnesium fluoride to multilayer broad-band designs (BBAR). If complex coatings are required, the user should know whether the supplier will fabricate the coating or subcontract to another shop.

While there has been great progress since Dobrowolski's review article, multilayer filters are still difficult to produce and prone to run-to-run variations and unexpected problems. Manufacturers can specialize in filter types and/or spectral regions. On-site visits may be necessary to determine whether a potential supplier has the facilities and personnel to undertake new and complex projects. On the other hand, there are cases where filter users should be willing to share risks by purchasing initial samples on a "best-efforts" basis.

The following sections present a variety of examples and include further remarks relevant to filter specifications. Curves were calculated and graphed with the FILM∗STAR design package running on an HP 9000 Model 320 computer and HP 7550 plotter. Data for absorbing glasses were obtained from tables included in the Schott catalog. Since these tables are only supplied to two significant figures, curves may include spurious minor bumps.

Most of the results can be reproduced with the supplied program. Unless otherwise indicated, unfilmed substrate surfaces do not include antireflection coatings. Two types of spectral curves are utilized: linear transmittance and logarithmic optical-density plots. While the former are useful for comparing edge behavior and optical performance in high-transmission zones, the latter are required for meaningful comparison in blocking regions.

Finally, it should be emphasized that all specifications should be as nonrestrictive as possible. Since there is no point in rejecting nonexisting light, pass bands and rejection zones need only cover wavelengths where system response is affected. This should be kept in mind through the remainder of this chapter.

6.3.1. Short-Pass Filters

The ideal short-pass (low-pass) filter transmits everything below and nothing above a given wavelength. Because glasses provide a broad cutoff region, multilayers must be utilized for sharp cut requirements. The fundamental dielectric design is[4]

$$\text{SUB} \quad [.125L \quad .25H \quad .125L]M \quad \text{AIR},$$

where L and H represent high- and low-index films such as quartz (SiO_2, $n = 1.46$) and titanium dioxide (TiO_2, $n = 2.3$), or magnesium fluoride (MgF_2, $n = 1.38$) and zinc sulfide (ZnS, $n = 2.3$). M is the repeat factor. That is

$$[.125L \quad .25H \quad .125L]3$$

is shorthand for

$$.125L \quad .25H \quad .125L \quad .125L \quad .25H \quad .125L \quad .125L \quad .25H \quad .125L$$

and equivalent to

$$.125L \quad .25H \quad .25L \quad .25H \quad .25L \quad .25H \quad .125L.$$

The filter has its transmittance minimum (reflectance peak) at the design wavelength where ".25" denotes quarter-wave optical thickness. Examples are shown in Fig. 3 for SiO_2 ($n = 1.46$) and TiO_2 ($n = 2.3$) deposited on a nonabsorbing glass substrate ($n = 1.52$). The design wavelength is 632 nm, which gives 50% cutoff at 550 nm for M = 10 (21 layers).

In most cases, constant indices n are adequate for calculations involving low-index dielectric films and glass substrates in the visible. While ignorable in the first approximation, dispersion should be included for high-index films—especially in the range of 400 to 500 nm. Note that TiO_2 and ZnS

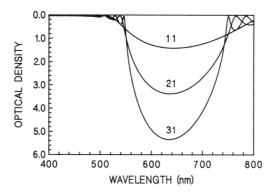

FIG. 3. Short-pass filter on glass—dependence on the number of layers.

become increasingly absorbing toward the ultraviolet. This can be detrimental to the performance of short-pass designs. To facilitate comparison with the results of the program FILTER, constant index values are utilized in this chapter (except for metals).

Performance in the passband can be improved by damped least-squares optimization[5] (refinement)—a method for automatically adjusting layer thicknesses according to desired optical performance. Using the same materials as in the previous example, typical improvements are shown in Fig. 4. Here the construction is

SUB .298H .284L [.25H .25L]8 .258H .280L .251H .138L AIR,

with design wavelength 628 nm. Because the substrate's second surface is uncoated, the maximum transmittance is 96%.

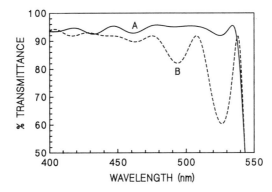

FIG. 4. A = Optimized short-pass design; B = Standard short-pass design.

Filter users whose critical requirements necessitate optimized designs should ensure that their supplier has access to appropriate thin-film design package and familiarity with methods for depositing nonquarter-wave layers. If a potential supplier cannot (which is often the case for new designs) provide measured curves showing the performance of previously manufactured samples, he ought to be able to supply computed linear and logarithmic spectral curves.

One difficulty with dielectric filters is the relatively narrow blocking zone. The zone can be widened by combining the above designs with one or more additional quarter-wave stacks centered at longer wavelengths. A typical design is

SUB [.125L .25H .125L]10 [.165L .33H .165L]10 AIR.

As this degrades transmission in the passband, optimized designs are almost certainly required. Further stacks become useless when the third-order reflectance peak (occurring at one-third the design wavelength) overlaps the passband. Such peaks do not occur on the long-wavelength side.

Short-pass filters constructed from filter glasses have a broad cutoff region and are often used in heat-absorbing applications. An example is Schott BG40. Spectral performance for various thicknesses are shown in Figs. 5 and 6. Sharp cutoff combined with broad rejection can be obtained by depositing a multilayer on filter glass (or by cementing two filters). This is illustrated in Fig. 7, where the 22-layer optimized short-pass filter has been combined with a 3-mm BG40.

6.3.2. Long-Pass Filters

The ideal long-pass (high-pass) filter transmits everything above and nothing below a given wavelength. Sharp cut devices can be constructed

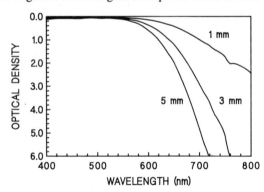

FIG. 5. Schott BG40. Optical density as a function of thickness.

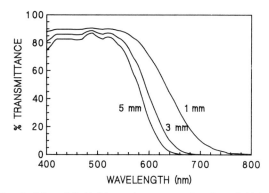

FIG. 6. Schott BG 40. Transmittance as a function of thickness.

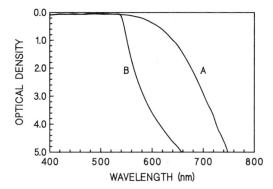

FIG. 7. A = 3-mm Schott BG40; B = Optimized short-pass filter deposited on 3-mm Schott BG40.

from multilayers or a wide choice of glasses. The basic long-pass construction,[4] using the same materials as in the short-pass example, is

SUB [.125H .25L .125H]M AIR.

A design wavelength of 475 nm results in 50% cuton at about 550 nm. Figure 8 illustrates dependence on layer number. The basic 21-layer design ($M = 10$) is compared to an optimized construction

SUB .106H .280L .185H [.25L .25H]8 .146L .356H .389L AIR

in Fig. 9. As before, the rejection zone can be widened by depositing additional quarter wave layers.

Colored glass filters are available for various cuton wavelengths. As an example, Fig. 10 illustrates spectral response for several thicknesses of

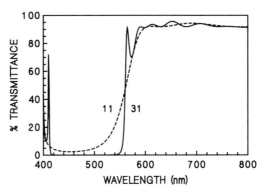

FIG. 8. Long-pass filter on glass.

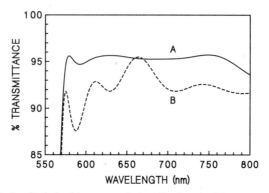

FIG. 9. A = Optimized long-pass design; B = Standard long-pass design.

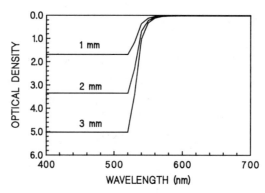

FIG. 10. Schott OG550. Optical density as a function of thickness. Density greater than shown from 400 to 500 nm.

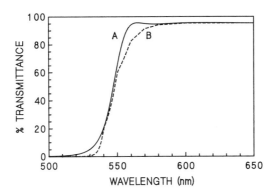

FIG. 11. A = Optimized long-pass design; B = 2-mm Schott OG550 with antireflection coating.

Schott OG550. The horizontal portions of the graph represent "less than" values. Figure 11 superimposes calculations for the 22-layer optimized multilayer long-pass and 2-mm OG 550. The slope for the multilayer can be increased by utilizing a design with quarter-wave layers. The filter glass calculation includes a four-layer broad-band antireflection coating applied to one surface (the other surface is uncoated). Note the slight differences in spectral sharpness near 0 and 100%.

6.3.3. Band-Pass Filters

The ideal band-pass filter transmits everything in a given wavelength range and nothing outside that range. Such isolating filters are often called *interference filters*. As a wide class of filters rely on interference, this is something of a misnomer. The band-pass filter is typically similar to a Fabry–Perot interferometer with the spacing between two high reflectors provided by a multiple half-wave dielectric spacer. Filters are constructed from all-dielectric or metal-dielectric multilayers. Whether one or the other is selected depends on bandwidth, blocking, and spectral range.

The optical performance of typical metal-dielectric filters is compared in Figs. 12 and 13. Three designs are evaluated:

1st Order   SUB   30M   202.7L   30M   AIR
2nd Order   SUB   30M   456.4L   30M   AIR
Augmented   SUB   30M   100.3L   153H   100.3L   30M   AIR

M indicates silver (dispersive), H is ZnS ($n = 2.3$), and L is MgF_2 ($n = 1.38$). Physical thicknesses are in nanometers. In the first-order design, the spacer layer is approximately an optical half-wave at the peak position. In the

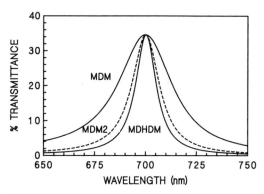

FIG. 12. MDM = first-order metal-dielectric-metal design; MDM2 = second-order design; MDHDM = augmented design.

second-order filter, the spacer is approximately twice as thick. Augmented filters[6] are those in which bandwidth is decreased through the addition of nonabsorbing half-wave layers to the basic construction.

Figure 12 illustrates the dependence of bandwidth on design. Second-order and augmented filters are clearly superior to first-order filters with respect to bandwidth. As shown in Fig. 13, this advantage is counterbalanced by decreased rejection on the long-wavelength side—especially for second-order filters. As the transmission of silver increases towards the blue, short wavelength blocking is required. Fortunately, a large selection of filter glasses is available for this purpose. The metal-dielectric can be directly deposited on filter glass or deposited on clear glass and cemented to the blocker.

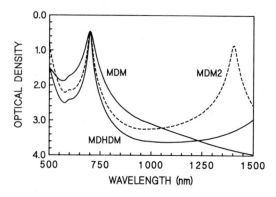

FIG. 13. Blocking characteristics of MDM filters.

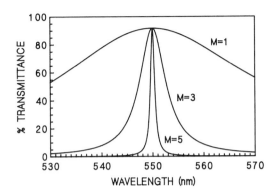

FIG. 14. Single-cavity dielectric band-pass design. 4M+3 layers.

All-dielectric filters can be manufactured with narrow bandwidth and high transmittance. A design for a first-order filter of this type is given by

SUB [.25H .25L]M .25H .5L .25H [.25L .25H]M AIR.

An alternative design uses a high-index spacer. Returning to the materials first utilized in Fig. 3 (L 1.46, H 2.3), the transmittance for this design is shown in Fig. 14. As the reflectance of the quarter-wave stack on either side of the .5L spacer increases with increasing M, improved bandwidth and rejection are obtained. Since the stack is "absentee" at design wavelength 550 nm, the maximum theoretical transmittance is independent of M.

The above design can be rewritten as

SUB C AIR,

where C stands for a single "cavity." N-tuple cavity filters are given by

SUB C [.25L C](N−1) AIR

Examples are shown in Fig. 15 for M = 3, N = 1, 2, 3, and in Fig. 16, for M = 2, N = 2. Once again both linear and logarithmic plots are required. The limited blocking region can be extended with filter glass on the low-wavelength side and broad-band multiple-stack dielectric coatings on the high-wavelength side. The second plot in Fig. 16 adds 2.5 mm Schott GG495 and a two-stack long-wavelength blocker optimized for transmittance from 500 to 600 nm. Blocking below 480 nm will actually be better than shown.

Costs can dictate the choice between an absorbing substrate coated on two sides or separate cemented substrates. Filter manufacturers can increase productivity by prefabricating long-wavelength blockers suitable for a range of filters. An alternative construction, especially suited to wide-band-pass

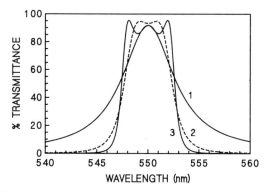

FIG. 15. 1-, 2-, and 3-cavity dielectric band-pass designs.

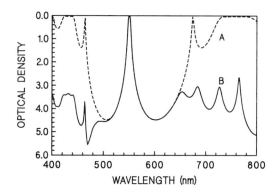

FIG. 16. A = 2-cavity dielectric band-pass design on glass; B = Same band-pass design plus short-pass on 2.5-mm Schott GG495.

requirements, uses long-pass and short-pass filters in series. This could include a single short-pass multilayer stack deposited on filter glass or, as shown in Fig. 17, short-pass and long-pass stacks deposited on opposite sides of the same substrate. If the stacks are deposited on separate substrates, each edge can be tuned (to shorter wavelengths) by tilting.

Returning to narrow-band-pass filters, bandwidths can be decreased by increasing the thickness of the spacer layer. Ultimately, thin films are no longer suitable and a solid spacer is required. Consider a nine-layer reflector deposited on both surfaces of a 2-mm glass substrate ($n = 1.52$) treated as a thin film. The design, centered at 550 nm is

AIR [.25H .25L]4 .25H 5527G .25H [.25L .25H]4 AIR.

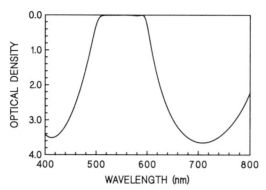

FIG. 17. Band-pass filter including optimized long- and short-pass designs on opposite sides of a glass substrate.

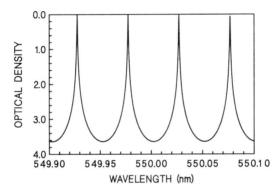

FIG. 18. 9-layer H-to-H reflector deposited on both sides of a 2-mm thick glass substrate. Coherent light assumed.

An optical density plot is given in Fig. 18. Unlike previous calculations involving reflection from substrate surfaces, amplitudes rather than intensities are summed. The filter is largely opaque with narrow transmittance fringes. This could be employed for the separation of narrowly spaced spectral lines. Intensity summation eliminates fringes and leads to the graph shown in Fig. 19. As these are identical constructions, which is correct? Should substrates be treated as "thin" or as "thick" elements?

In examining the fringe spacing shown in Fig. 18, it can be deduced that exceptionally flat substrates are required. As substrates deviate from perfect flatness, amplitude addition is replaced by intensity addition: fringes become eliminated. This can be verified through repeated thin-film calculations by assuming a distribution in substrate thickness. As a further requirement for observing fringes, the analyzing instrument must be highly monochromatic.

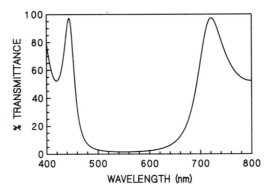

FIG. 19. Same construction as Fig. 18. Incoherent light assumed.

As bandwidth is increased, the incoherent (intensity addition) result is obtained in the limit. While fringes are not normally observed in typical applications involving optical filters, caution should be exercised when using multilayers with laser radiation. Because the substrate need only be flat over a narrow beam width, undesired interference effects could occur.

6.3.4. Rejection Filters

The ideal rejection filter rejects all light in a given spectral range and transmits everything else within the zone of interest. Quarter-wave stacks are frequently utilized in such applications. A basic design is

SUB [.25H .25L]M .25H AIR.

As shown in Fig. 20 for M = 10, rejection and bandwidth depend on film materials. The low-index film is SiO_2 (1.46); high-index materials are Al_2O_3

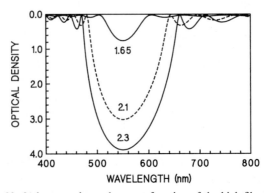

FIG. 20. 21-layer stack on glass as a function of the high film index.

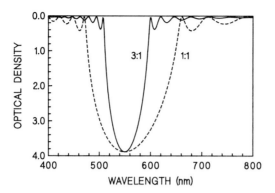

FIG. 21. Comparison of 1:1 and 3:1 (.75H/.25L) 21-layer stacks on glass.

(1.65), Ta_2O_5 (2.1), and TiO_2 (2.3). As illustrated previously in Fig. 3, slope and minimum transmittance also depend on the number of layers.

Further spectral variations are possible with designs containing multiple quarter waves. Figure 21 compares the previous design (for SiO_2, TiO_2, $M = 10$, 550 nm) with a design including triple quarter-wave layers given by

SUB [.75H .25L]10 .75H AIR.

One problem with such designs is the appearance of undesirable rejection zones at shorter wavelengths. As illustrated in Fig. 22, optimization provides another possibility for narrowing the normal rejection zone of a TiO_2–SiO_2 quarter-wave stack deposited on glass (1.52). The design is given by

SUB .406H .167L .057H .347L [.310H .311L .132H .225L]3 .435H .098L .179H .590L AIR,

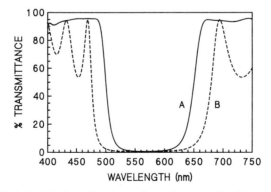

FIG. 22. A = Optimized design with narrowed rejection zone; B = Quarter-wave stack using same film materials.

with design wavelength at 575 nm. Compare this to the superimposed spectrum for

$$\text{SUB} \quad [.25\text{H} \quad .25\text{L}]8 \quad \text{AIR}$$

(design wavelength 560 nm). While we offer no theoretical justification, it appears that the central group in the optimized design simulates a quarter-wave stack of materials with indices closer than actual indices. The outer layers supply short-pass and long-pass antireflectance.

There is interest in filters which reject in extremely narrow spectral regions. These could be superior to current filters utilized for eye protection near lasers. In theory, such filters can be constructed from a great number of quarter-wave layers of materials with nearly identical indices. Inhomogeneous films with periodic index variations represent another possibility. As new deposition techniques are developed, fabrication of such filters becomes increasingly feasible.

6.3.5. Color-Correcting Filters

Neutral-density filters are used to reduce the amount of light without affecting relative spectral composition. A variety of filter glasses are suitable for this purpose. If reflected light is acceptable (or desirable), neutral-density filters can be constructed from vacuum-deposited thin-metal films, especially nichrome. Through an assortment of rotating masks and other proprietary deposition fixtures, metal films can be deposited in a linear or circular wedge. The net result is the extremely useful variable-transmittance filter. While dielectric multilayers could be used as neutral-density filters, their higher cost requires special justification.

Colorimetry[7] is the branch of optics which relates spectral composition to our perception of color. Color may be specified according to chromaticity coordinates calculated from CIE tristimulus (blue, green, red) values from 380 to 770 nm at 5-nm intervals. When plotted on an xy scale ($0 < x < 1$, $0 < y < 1$), the locus of colors corresponding to blackbody sources is called the *Planckian locus*. "Standard illuminants" include type A (tungsten lamp at 2854° K), type B (sunlight at about 4880° K), and type C (daylight at about 6740° K).

Color-correcting filters change the balance of energy in the visible and are used in a variety of applications. Photographers rely on such filters for converting available light to match the characteristics of color film, and a variety of color-balancing filters (with durable broad-band AR coatings) can be found in camera shops.

Multilayers are appropriate when requirements cannot be met with colored glasses. As an example, consider an all-dielectric filter for converting

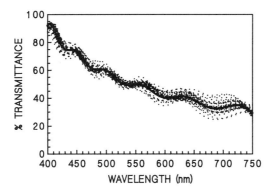

FIG. 23. 3300 to 4200° K filter design (solid) showing effect of random thickness errors (dotted).

light from 3300 to 4200° K. Using practical materials and optimization targets calculated from blackbody intensity ratios, the following design is obtained:

SUB .258H .217L .149H .123L .242H .242L .278H
.260L .267H .289L .272H .309L .264H .153L AIR.

Here L is SiO_2 (1.46), H is Ta_2O_5 (2.1), and SUB is glass (1.52); the design wavelength is 700 nm. The filter's spectral response is illustrated by the solid line in Fig. 23. The trichromatic coefficients are $x = 0.372$ and $y = 0.371$ corresponding to approximately 4208° K. Luminance, an indication of total transmitted light, is $Y = 0.468$ ($Y = 1.0$ for $T = 100\%$).

A color-correction filter meets standards if its xy coordinates fall within an acceptable range on a chromaticity plot. The fabricator can utilize the same plot to estimate manufacturing errors through repeated calculations with randomly varying film thicknesses. Using the color-correcting design given above, spectral variations for optical-thickness errors with standard deviation 0.005 are shown as dotted lines in Fig. 23. Figure 24 displays the same variations on a chromaticity plot. Final yield and price depend on the range of values deemed acceptable.

The above provides an example of filter specifications which are based on computations involving transmittance values—not on transmittance values themselves. Ultimately the user will want to illuminate the filter and measure the color temperature of transmitted light. Great care must be taken to ensure the repeatability of the light source as well as the accuracy of the device utilized for determining chromaticity coordinates.

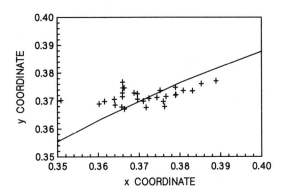

FIG. 24. Results of Fig. 23 plotted on CIE chromaticity coordinates.

6.3.6. Reflectance Devices

A number of useful devices can be constructed from multilayer reflectors. Here reflected light is retained and transmitted light eliminated from the final beam. The "cold mirror," a popular example, is a long-pass design combining two or three adjacent quarter-wave stacks for high reflectance across the visible. Infrared energy passes through the substrate and is thereby eliminated from the reflected beam. Modern projection lamps and medical illuminators rely on cold mirrors.

Reflector assemblies can replace long-wavelength blockers in band-pass designs. For example, the optimized long-pass design illustrated in Fig. 9 reflects light from about 420 to 540 nm. Suppose that two such filters are arranged in a multiple reflectance configuration, as shown in Fig. 25. Assume that the back surfaces are ground to eliminate unwanted reflections. Assuming near-normal incidence, the spectral response of this assembly (plotted as optical density) is shown in Fig. 26.

The performance of such devices can be quite superior to multiple stack blockers. It would seem that far lower coating costs (22 layers in the visible as compared to 60 layers or more in the infrared where coatings are much thicker) could offset increased fixturing costs. There are numerous other

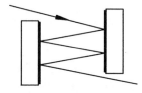

FIG. 25. 4-bounce long-wavelength rejection filter.

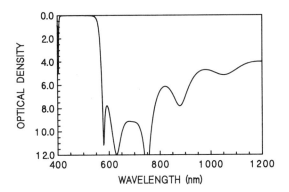

FIG. 26. Optimized long-pass design used as a 4-bounce long-wavelength rejection filter (see Fig. 25).

instances where similar devices would simplify construction and improve optical performance.

If reflectance devices offer numerous potential advantages, why are they not utilized more widely? Probably because filters are often added to existing systems rather than included as integral components during system layout. Optical-system engineers may be unfamiliar with filter properties and lack computational tools for their evaluation. Furthermore, planar transmitting elements are compact and can easily be inserted in various locations in the optical path of existing systems. In the case of non-normal incidence, reflectance devices could also introduce undesired polarization effects.

6.4. Selecting Filters: Secondary Criteria

While spectral response is the initial consideration when selecting filters, there are other important criteria. Multilayer devices reject light by reflection (and absorption if absorbing films are included in the design). This could result in undesired spectral energy elsewhere in the system. Because the transmittance of any filter-substrate assembly is independent of the direction of incident light, unwanted reflection might be minimized by simply flipping the filter in the beam. System designers should also be aware that filter-glass fluorescence is another possible source of unwanted light.

The spectral properties of colored glasses shift with temperature. Although catalogs provide relevant coefficients, it may be necessary to measure samples at different temperatures. Multilayer devices may be subject to spectral instabilities. Filters with nearly identical spectral response can be constructed from "hard" or "soft" materials with vastly differing aging characteristics. Stability and durability are increased by cementing a cover glass over the multilayer filter. In critical cases, designers should include temperature and humidity criteria in their specifications.

Spectral shifts will result from non-zero angles of incidence. Multilayer spectral features shift to shorter wavelengths—an effect which can be used to advantage. If the incident beam subtends an angular range, bandwidths are increased and sharp cutoffs broadened. Similar effects are observed when spectrally sharp devices are measured with light of insufficient monochromaticity. Absorbing filters appear "thicker" when tilted—a consequence of the longer light path predicted by Snell's law. Such effects are calculable and within the expertise of filter manufacturers having access to appropriate design software.

6.5. Measuring Filter Performance

Users and manufacturers must agree on methods for determining whether finished filters meet specifications. In most cases, these measurements can be successfully performed with commercial spectrophotometers such as those supplied by Perkin-Elmer and Varian. There are other cases where fabricators must construct special apparatus.

Extra care is required in a number of instances: (1) The accuracy of transmission readings in blocking regions will depend on the instrument's stray-light rejection. Doubly-dispersing instruments may be required. (2) The bandwidth of measuring light must be narrower than the spectral features of the filter being measured. Opening monochromator slits for better signal-to-noise can result in highly erroneous readings. (3) Filters measured at non-normal angles may require knowledge of the instrument's polarization characteristics.

In the case of reflectance specifications (typically for AR coatings), there is a choice: (1) Compare uncoated and coated samples of the same filter; calculate reflectance from transmittance changes. This works best for nonabsorbing filters; the fabricator's filter-design software can be utilized to ascertain the decreasing accuracy of this method for increasingly absorbing materials. (2) Directly measure the reflectance. This is usually more difficult and more expensive than transmittance measurements. System designers should not specify reflectance criteria when they are really interested in maximizing transmittance.

Newer spectrophotometers such as the Varian 2300 can be purchased with IEEE-488 or RS-232C computer interfaces. Desktop computers can be used to control, acquire, and manipulate data, and determine whether a filter meets specifications. By utilizing integrated design and measurement software, fabricators using such apparatus will be in a better position to supply filters that meet complex spectral requirements. Filter fabricators should be wary of standard computer hardware/software packages provided by spectrophotometer suppliers. These are usually intended for chemists and may provide little support for optical films.

Spectrophotometers are generally used in continuous scan mode. This is appropriate when the output consists of strip-chart traces. As computer analysis requires measurements at discrete wavelengths, the following sequence is suggested: rapid scan to next wavelength, pause according to time constant (and/or other criteria), data acquisition. Since CIE analysis, for example, only requires measurements at 5- or 10-nm intervals, considerable time may be gained.

6.6. Vacuum Deposition Techniques

We assume that the reader intends to procure filters from a supplier within his organization or from an external source. If the devices are similar to those previously supplied, it would seem a simple matter to judge whether the manufacturer can meet current requirements. Nevertheless, caution is always in order. A coating shop that has successfully produced 50 filters on 1.25 inch diameter substrates may not be able to meet requirements for 1000 filters on 12 inch substrates.

If the filter is outside the supplier's past experience, the user must determine whether that supplier has sufficient skills and equipment for the task. In addition to designing and measuring the product (previously discussed), the supplier will need to fabricate substrates and deposit multilayer coatings. In the case of filters, substrates present few difficulties. In nearly all cases, problems involve coatings.

Nearly all multilayer coatings are fabricated in high-vacuum chambers via thermal evaporation. In most cases, vacuum is maintained by a diffusion pump with a cooled baffle mounted between the pump and chamber. If sufficient cooling is provided, the baffle itself will remove water vapor—a major contaminant in most systems. The diffusion pump may be altogether eliminated in favor of completely cryo-pumped systems.

As mentioned before, coatings may be fabricated from "hard" (SiO_2, TiO_2) or "soft" (MgF_2, ZnS) materials. Hard coatings are deposited in a chamber equipped with an electron-beam gun. Substrates are heated to 250°–350°C, and the chamber is backfilled with oxygen during evaporation of the high-index layers. Soft coatings (which may be quite durable!) are applied at somewhat lower temperatures to prevent reevaporation of ZnS. Here dielectrics are heated with resistive elements—either radiantly from a "pancake" filament or within a "howitzer," or by placing the material in a metallic "boat."

Should hard or soft coatings be specified? Soft coatings can be chemically removed, while hard coatings require repolishing. If the hard-coating process is known to have a 75% yield, the filter user should be willing to pay for extra substrates or be prepared to wait for repolishing. The lesser environmental stability of soft coatings could be of minimal consequence if the filmed surface is sealed with a cemented cover glass. In some cases, filter manufacturers will claim "trade secret" and refuse to reveal their choice of thin-film materials. Filter purchasers should be warned that soft coatings often include radioactive thorium compounds. While such films may only be dangerous if eaten, there may be regulations which proscribe their use.

Two vital considerations for the production of optical films are uniformity and control of layer thickness. If only a few substrates are to be coated, or if uniformity is not critical, a dome-shaped stationary fixture may be sufficient. More exacting uniformity specifications are met with a rotating fixture, and it is generally agreed that a planetary (two degrees of rotation) setup with stationary or rotating coating masks is required in order to achieve the most critical uniformity requirements.

The required accuracy in controlling layer thickness depends on the particular coating design. Relatively crude monitoring is sufficient for blocking filters, whereas a high degree of control is required for band-pass filters. The problem is compounded by the need to achieve repeatable refractive indices.

In most instances, direct or indirect optical monitoring is employed for thickness control. In direct monitoring, the optical transmittance or reflectance of the actual filter is measured at a selected monitoring wavelength during deposition; separate witness samples are employed for indirect monitoring. While indirect monitoring can be utilized with high-capacity planetary fixturing, there are cases where direct monitoring results in higher yields.

In either case, a computer program should be available for translating design information into manufacturing parameters. Coating shops without such programs may be utilizing less than optimum optical-monitoring strategies. Manufacturers who restrict optical monitoring to integral quarter-waves may be unable to fabricate optimized designs.

In recent years, there have been advances in the "crystal" monitor—a device based on the change in resonant frequency of a quartz crystal as its mass is increased by evaporated material. While this device is generally accepted as suitable for controlling the rate of deposition of optical materials, there is disagreement on whether it is sufficiently accurate for thickness control. Optical deposition chambers are often configured to include both types of monitoring.

Using results from previous deposition runs, coating engineers can employ random film analysis to estimate standard deviations in layer thicknesses of each material. This can, in turn, be applied to new designs, as in the previous example of a color-correcting filter. A user faced with widely divergent cost estimates for the same filter must determine whether the low bidder is actually more competent or whether he is unaware of the difficulties.

6.7. Discussion and Conclusion

This chapter has been intended for those who require filters and are interested in the general characteristics of the various types used in transmission.

Filter design and selection are aided by powerful commercially available thin-film design programs running on desktop computers interfaced to spectrophotometers. Fabricators with appropriate hardware and software should be able to meet new and complex requirements.

The situation in multilayer design is rapidly evolving as new deposition technologies, such as those involving ion-assisted evaporation and sputtering, increase the total number of layers which may be deposited. New types of filters will arise from the freedom to utilize hundreds of thin films in a single design. In turn, this will place increased demand on hardware and software for designing and measuring filters.

In most cases, filters are added to previously designed systems. As computers become more powerful, it will be increasingly possible to integrate optical filters, including absorbing multilayers utilized in reflectance, into system designs.

References

1. J. A. Dobrowolski, "Coatings and Filters", in *Handbook of Optics*, (W. G. Driscoll, ed.) Chap. 8, McGraw-Hill, New York, 1978.
2. H. A. Macleod, *Thin Film Optical Filters*, 2nd Edition, Macmillan, New York, 1986.
3. F. T. Goldstein, FTG Software, P.O. Box 358, Chatham, NJ 07928.
4. L. I. Epstein, "The Design of Optical Filters," *J. Opt. Soc. Am.* **42**, 806 (1952).
5. M. J. Kidger and C. G. Wynne, "Experiments With Lens Optimization Procedures," *Optica Acta* **14**, 279 (1967).
6. A. F. Turner, "Some Current Developments in Multilayer Optical Films," *J. Phys. Radium* **11**, 444 (1950).
7. D. L. MacAdam, "Colorimetry," in *Handbook of Optics* (W. G. Driscoll, ed.), McGraw-Hill, New York, 1978.

INDEX

A

Abbe prism, 66
Abbe refractometer, 187
Abbe sine condition, 24
aberrated wavefront, 44
aberration polynomial, 40, 44
achromatic objectives, 164
adjustment distance, 165
afocal systems, 132, 138
Airy disc, 73, 168
Amici lens, 129
Amici prism, 66, 187, 210
angular dispersion, 200
angular field of view, 131
angular magnification, 133
apertometer, 166
apherical surfaces, 7, 53, 94
aplanatic lens, 110
apochromatic objectives, 164
apodization, 226
apparent field, 61
arc
    carbon, 249
    compact source, 247, 249
astigmatism, 42
autocollimator, 142

B

band-pass filters, 287
base
    candelabra, 242
    mogul, 242
    prefocus, 242
basic laws of geometrical optics, 10
beam expanders, 133
Becke line, 189
Bertrand lens, 173
binocular microscope, 170
binocular telescope, 59
black-light lamp, 252
blazed grating, 214
brightness, 195

bulb lamp, 252
    parabolic, 240
    reflector, 240
    tubular, 240
Burris diode, 256

C

calculation sheets for thin lenses, 29
camera lenses, 74, 95
carbon arc sources, 248
cardinal points of a lens, 25
catoptric optics, 169
center of perspective, 80
characteristic matrix for thin films, 274
chief ray, 37
chromatic aberrations, 30
Coddington's equations, 17
coefficient of finesse, 218
coherence length, 258
cold mirror, 296
collection power, 195
color correcting filters, 294
coma, 42
compact source arc, 249
    Mercury, 250
    Mercury-Xenon, 250
    Xenon, 250
compensating eyepiece, 132
compound microscope, 159
concave grating, 215
condensers, 99
confocal scanning microscope, 174
conic constant, 53
conoscopic mode, 173
constant-deviation spectrometers, 208
critical illumination, 152
critical-angle refractometers, 187

D

dark field illumination, 172
depth of field, 167
design concepts, 92

design tips, 51
detection of objects, 124
differential refractometer, 185
diffraction efficiency, 214
diffraction limited lens, 166
diffuse illumination, 153
dipping refractometer, 188
direct viewing prism, 210
discharge sources, 247, 251
dispersion spectrometers, 200
distortion, 42
do-it-yourself optical engineers, 3
double Gauss lens, 84
Duc de Chalnes method, 190
dye lasers, 263

E

Ebert-Fastie arrangement, 216
ellipsometer, 179
empty magnification, 168
Erfle eyepiece, 62
etalon, 221
etendue, 195
exit pupil, 168
eye, 120
  circle, 131
  relief, 61, 131
eyepieces, 59, 71, 127, 129, 131

F

F/number, 25
Fabry-Perot, spherical, 220
Fabry-Perot
  etalon, 191, 217
  spectrometers, 217
far point of the eye, 120
far sighted, 68
Fellgett advantage, 198
field lens, 130
field stop, 37, 131, 171
field-widened interferometer, 227
filaments, 240
  coiled, 240, 244
  emissivity of, 240
  ribbon, 240
filar micrometer eyepiece, 132
filling gas, 241
filter
  performance, 289
  program, 276

filters, 282
  optical, 273
filtrage, 196
finite conjugate lenses, 91
flash-discharge lamp, 254
fluorescence microscopy, 174
fluorescent lamps, 251
  instant start, 252
  preheated, 252
  rapid start, 252
fluorite, 36
focal shift, 42
focal systems, 132
Focault knife edge test, 175
Fourier components, 156, 167
Fourier transform, 148,
Fourier transform spectrometer, 190, 222
fractional height, 39
free spectral range, 198
free working distance, 165
future in optics, 5

G

Galilean telescope, 69, 136
galvanometer mirror scanners, 113
gas diode laser, 267
gas lasers, 262
Gauss equation, 17
Gaussian beam, 107, 260
Gaussian optics, 22
germicidal lamp, 253
Goulay method, 206
graphical ray tracing, 18
graticules, 131
grating equation, 210
grating spectrometers, 210

H

half-shade device, 124
heterojunction laser, 268
Hilger-Chance refractometer, 185
holographic scanners, 114
holography, 266
homojunction laser, 268
Huygens eyepiece, 61, 132

I

illuminance, 100, 104
illumination, 146
illumination systems, 106, 170

image power, 155
immersion magnifier, 129
immersion objectives, 168
immersion refractometer, 188
infinity-corrected objectives, 165
infrared microscope, 174
instantaneous field of view, 139
instrument function, 195
instrument profile, 195
instrumentation lamp, 244
interferogram, 224
interferogram function, 225
intrinsic resolving power, 197
inverted telephoto, 89

J

Jacquinot advantage, 198
Jelley refractometer, 190

K

Kellner eyepiece, 62
Keplerian telescope, 135
keystone distortion, 138
knife edge, 175, 178
Koheler illumination, 99, 152

L

Lambert's law, 240
lambertian source, 99
lambertian surface, 99
landscape lens, 75
Langrange equation, 22
laser beam divergence, 260, 261
laser beam waist, 260, 261
laser scanners, 112, 114
laser sources, 258
laser systems, 106
Laser
  Neodymium YAG, 265, 267
  Neodymium glass, 265
  Rubi, 265
  dye, 263
  gas, 262
  gas diode, 267
  semiconductor, 265, 267
  solid state, 265
law of refraction, 10
LEDs, 254

lead-in wires, 241
length standards, 253
lenses for laser scanners, 114
light integrating bar, 103
light emitting diodes, 254
light enhancement factor, 258
light gathering power, 195
light integrating bar, 103
Littrow arrangement, 201, 213
long-pass filters, 284
luminance, 100
luminosity, 195
Lummer-Brodhun cube, 123

M

magnifier, 67, 127
Maxwell equations, 274
mechanical tube length, 165
mechanical zoom, 96
meridional plane, 7
meridional ray tracing, 14
metascope, 9
Michelson interferometer, 223
microdensitometer, 174
microphotography, 159
microscope, 67, 159
  eyepieces, 71
  objectives, 164
  physical parameters, 162
mirror systems, 21, 97
mirror, cold, 296
modes, 259
modulation lamp, 244
modulation transfer function (MTF), 76
monochromator, 191
Mueller matrix ellipsometers, 181
multilayer thin films, 274
multimode laser, 260
multiple pass spectrometers, 201
multiplex advantage, 198
multiplex factor, 198

N

near point of the eye, 120
nearsightedness, 68
Nelsonian illumination, 152
Neodymium
  YAG laser, 265
  glass laser, 265, 267

Newton relation, 28
nodal planes, 25
normal dispersion, 201
normal magnifying power, 70
notation in geometrical optics, 13
numerical aperture, 24, 29, 166

O

off-the-shelf components, 3
optical
  axis, 7
  components, 59
  design, 4
  design by computer, 51
  filters, 273
  instruments, 119
  methods, 1
  path (OP), 11
  path difference, 11, 44
  sine condition, 24
  tube length, 165
  zoom, 96
order of interference, 191, 210
order sorting, 213
orthoscopic
  eyepiece, 62
  mode, 173

P

paraxial computation system, 19
paraxial ray, 17, 39
parfocal distance, 165
partial coherence, 109, 146
peak-to-valley, 106
Pechan prism, 66
Pechan-Schmidt prism, 64
Pellin-Broca prism, 208
pentaprism, 66
Petzval
  coefficient, 42
  curvature, 42
  lens, 90
phase contrast, 173
photoflash, lamp, 247
photolithography, 138
photomicrography, 159
plano objectives, 165
polar nephelometers, 181
polarization microscope, 173

polygon scanners, 113
Porro prisms, 64, 137
power throughput, 195
pressure scanning, 222
principal planes, 23, 25
principal refracting plane, 25
prism goniometer, 188
prism spectrometers, 206
prisms, 64
projection lenses, 99
projectors, 143
Pulfrich refractometer, 188
pupils, 37

Q

quartz, 36

R

Ramsden circle, 131
Ramsden eyepiece, 62, 132
ray tracing, 10, 53
Rayleigh
  criterion, 166, 303   207
  criterion for a prism, 207
  refractometer, 190
rays, a and b, 22
ready-made optics, 3
recognition of objects, 124
reflective
  devices, 296
  optics, 169
refractometers, 183
rejection filters, 292
relay unit, 137
residual color, 34
resolving power, 196
reticles, 131
retinoscopic lamp, 244
Rheinberg illumination, 173
Ronchi ruling, 159
roof Prism, 65
Rowland circle, 215
Rubi laser, 265

S

saggital
  astigmatism, 42
  coma, 42

INDEX

scanners, 113, 114, 140
scanning etalons, 222
scanning microscope, 174
schlieren instruments, 175
secondary color, 34
semiconductor laser, 265, 267
shadow test, 179
short-pass filters, 282
shoulder height, 165
sign convention, 14
single frequency laser, 260
Snell's Law, 11
solid state lasers, 265
sources filament, 239
Sparrow resolution criterion, 218
spectral
    lamp, 253
    purity, 196
    resolution, 196
    unit, 190
spectrograph, 191
spectrometer, 191, 200
spectrometer transfer function, 196
spectrophotometer, 191
spectroradiometer, 191
spectroscope, 191
spectroscopic instruments, 190
spherical Fabry-Perot, 220
spherical aberration, 42
spherocylindrical surfaces, 56
stage microscope, 159
standard irradiance, 247
Stokes vector, 181
stops, 37
substage condenser diaphragm, 171
sun, 239
support wires, 242
surface contributions, 46
symmetrical eyepiece, 62
system magnification, 24

T

Tau Law, 61
telecentricity, 37, 157
telephoto lenses, 86
telescope, 30, 59, 132, 135
    objectives, 64
terrestrial telescope, 36, 59, 136
thin films, 274

thin lenses, 28
third order
    aberration coefficients, 46
    aberrations, 40
    spherical aberration, 42
toric surface, 8
transverse axial chromatic aberration, 32
transverse electric mode, 261, 266
transverse magnification, 133
transverse rays plots, 43
triplet lens, 79
tube lens, 165
tungsten evaporation, 244
tungsten halogen lamp, 246
tungsten lamp, 239
Twyman-Green interferometer, 223
types of optical systems, 7

U

ultramicroscopy, 173
ultraviolet microscope, 174

V

vacuum deposition techniques, 299
vee-block refractometer, 185
vernier acuity, 124
vignetting, 37, 130
visual acuity, 124
visual system, 120

W

Wadsworth arrangement, 208, 216
wave aberration polynomial, 42
Weber law, 123
Wood anomalies, 214

Y

YAG laser, 254
YNU, method of paraxial ray tracing, 19

Z

z-system, 177
zoom eyepieces, 132
zoom lenses, 95